Successful Product and Business Development

Edited by
NEWMAN H. GIRAGOSIAN

President
Delphi Marketing Services, Inc.
New York, New York

MARCEL DEKKER, INC. New York and Basel

The opinions expressed in this book are those of the authors and do not necessarily reflect the views of the Commercial Development Association, Inc. Anyone making use of this book assumes all risk and liability arising from such use.

Library of Congress Cataloging in Publication Data

Main entry under title:

Successful product and business development.

Supported by the Commercial Development Association.
Bibliography: p.
Includes index.
1. Business–Addresses, essays, lectures.
2. New Products–Addresses, essays, lectures.
I. Giragosian, Newman H. II. Commercial Development Association.
HF5351.S84 658.5'75 78-10158
ISBN 0-8247-6770-5

MARCEL DEKKER, INC.

270 Madison Avenue, New York, New York 10016

Current printing (last digit):

10 9 8 7 6 5 4 3 2 1

PRINTED IN THE UNITED STATES OF AMERICA

FOREWORD

This book follows by more than 20 years the Commercial Development Association's invaluable text *Successful Commercial Chemical Development.* In the intervening years many articles have considered various aspects of the commercial development function, and several books have included commercial development as an element of much broader treatments of the total marketing function. During this same period, however, many changes have occurred in the framework and nature of commercial development: whereas the fundamental role of the commercial development function remains essentially unchanged, the structure, methods, and sometimes even the name of the function have changed in keeping with changes in the business environment in which it operates. As part of the change the Commercial Chemical Development Association has recently changed its name to the Commercial Development Association, acknowledging the broadening viewpoint and scope of this profession.

This book was conceived in the recognition of the changes which have occurred and which are occurring in commercial development and in the business environment. Its editor, Newt Giragosian, has successfully and painstakingly assembled pertinent chapters and case studies on key commercial development functions as now practiced. Some of the contributors to the current undertaking were active in commercial development when CDA's original book, edited by Hoyt Corley, was written; others are newer to the field.

Taken together, the committee members and contributors to this volume represent the widest possible range of commercial development experience, in terms of length and depth of service, to a broad spectrum of industry sectors.

Perhaps the firmest conclusion to be drawn from even a casual reading of this book is that commercial development remains a dynamic and vital function of all companies concerned with their future prospects. Under whatever name or label commercial development is practiced, it remains in the vanguard of progressive corporate activity; the passage of time and changing circumstances have enhanced and not diminished its importance.

We are deeply indebted not only to the authors who have contributed to the various chapters of this book, but also to the committee and its editor, Dr. Newman H. Giragosian.

Henry Avery

PREFACE

In a very real sense, *Successful Product and Business Development* was born in 1969, when the Board of Directors of the Commercial Development Association voted that a comprehensive book describing the current state and future prospects for commercial development was needed, and that such an undertaking should be supported by the association. This decision recognized the changes that had taken place both in the scope and nature of the commercial development function and in the character of the association membership: since the publication of *Successful Commercial Chemical Development* in 1954, many changes had taken place that had altered the concept and methodology of chemical commercial development. With the blurring of the lines dividing the chemical industry from other industries, and with the forward and backward integration of the old-line chemical companies, the world of the commercial development function had been changed considerably. There were seven primary reasons for this:

1. Entry into the chemical industry by nonchemical companies and the acquisition of nonchemical companies by chemical companies
2. Rising obsolescence rate of products
3. Declining profits
4. Rising operating costs and plant investment costs
5. Price weakness combined with intra- and interindustry competition
6. Declining productivity of research and development, combined with rapidly rising costs of research and development
7. Federal and state government action concerning resource pollution, occupational safety, tariffs, patents, and taxation

Some of these forces merely changed the way commercial development was being accomplished. Others, however, forced a reevaluation of the scope of the function and its place in the company's organizational structure and a critical review of the usefulness of having formalized groups carry out this type of activity.

This book attempts to delineate these changes and, at the same time, show how the people carrying out industrial commercial development today perform this vital function.

The title of the book, *Successful Product and Business Development,* reflects one important change that has occurred during the past two decades; that is, the change in thinking of chemical companies from a product-oriented outlook to a business-oriented one. This book attempts to reflect this change and, at the same time, recognize the often-forgotten fact that businesses are made up of products.

This book has been organized into five sections:

1. The introduction (Part I), which deals with the organization and structure of commercial development
2. Managing the commercial development function, in which tools and methodology, selection and evaluation of new products, and pricing are discussed (Part II)
3. The relationship of commercial development to other corporate functions (Part III)
4. Commercial development in its international setting and in terms of future possibilities (Part IV)
5. Case histories (Part V)

In choosing a book committee to guide and direct the synthesis of *Successful Product and Business Development,* great care was taken to enlist the active support of senior commercial development managers and practitioners from a broad range of company sizes, types, and viewpoints. The committee, in its turn, was able to enlist the advice, opinion, and aid of a great number of commercial development personnel throughout industry. As a result, the book represents the combined input of a broad cross section of the commercial development community, whose contributions are gratefully acknowledged; these include Ralph Herbst and Hal Johnson. Special appreciation is also due to the chapter authors and to the book committee—Henry Avery, J. Kenneth Craver, Aimison Jonnard, William Kinsinger, Edwin S. Michaels, and Carl Pacifico.

Newman H. Giragosian

CONTRIBUTORS

Affleck, James G., Chairman and President, American Cyanamid Company, Wayne, New Jersey

*Alcorn, David S.,** President, Dyes and Chemicals Division, Crompton & Knowles Corporation, Fair Lawn, New Jersey

Avery, Henry, Vice-President, USS Chemicals, Division of United States Steel Corporation, Pittsburgh, Pennsylvania

Copulsky, William, Office for Operations, W. R. Grace & Company, New York, New York

Giragosian, Newman H., President, Delphi Marketing Services, Inc., New York, New York

Gisser, Philip, Foster D. Snell Division, Booz-Allen & Hamilton Inc., Florham Park, New Jersey

Leenhouts, James W., Manager, Business Development, Michigan Division, Dow Chemical U.S.A., Midland, Michigan

McNeil, J. William, President, McNeil Associates, Pittsburgh, Pennsylvania

Michaels, Edwin S., Director of Development, Specialty Chemical Division, Stauffer Chemical Company, Westport, Connecticut

*Affiliation at time of writing: Executive Vice-President, J. T. Baker Chemical Co., Phillipsburg, New Jersey.

Pacifico, Carl, Management Supplements, Columbia, Maryland

Polacek, Robert J., Management Supplements, Briarcliff Manor, New York

Sharrard, George F., Kennecott Copper Corporation, Lexington, Massachusetts

Tinsley, Samuel W., Director, Corporate Technology, Union Carbide Corporation, New York, New York

Trebellas, John C., Technical Manager, Celanese Chemical Company, Inc., Technical Center, Corpus Christi, Texas

Vona, Joseph A., Director of Marketing–Technical Development Laboratory, Celanese Chemical Company, Inc., Summit, New Jersey

CONTENTS

Foreword iii

Preface v

Contributors vii

PART I Introduction and Scope of Commercial Development 1

1 Past, Present, and Future of Commercial Development 3
Carl Pacifico

2 Organizing for and Managing Commercial Development 15
James W. Leenhouts and Samuel W. Tinsley

3 Innovation and Creation of New Product Ideas 33
Newman H. Giragosian

4 Business Planning 43
Henry Avery and J. William McNeil

5 New Venture Management 61
William Copulsky

6 Inputs to Commercial Development 75
Newman H. Giragosian, Edwin S. Michaels, Joseph A. Vona, and John C. Trebellas

PART II Managing the Function 97

7 Understanding the Environment of New Business Ventures 99
George F. Sharrard

8 The Selection and Evaluation of New Projects 115
Newman H. Giragosian and Carl Pacifico

9 New Product Pricing 143
Newman H. Giragosian

10 The Communications Program 157
Philip Gisser

11 External Sources of Commercial Development Assistance 181
Newman H. Giragosian

PART III Interrelationships 205

12 What Top Management Expects from Commercial Development 207
James G. Affleck

13 Interrelationships with Other Corporate Functions 217
David S. Alcorn

PART IV Other Aspects of Commercial Development 229

14 International Commercial Development 231
Robert J. Polacek

15 Future Trends in Commercial Development 245
Newman H. Giragosian and Robert J. Polacek

PART V Selected Case Histories of Commercial Development Projects 251

Bibliography 291

Index 299

PART I

INTRODUCTION AND SCOPE OF COMMERCIAL DEVELOPMENT

1
PAST, PRESENT, AND FUTURE OF COMMERCIAL DEVELOPMENT

Carl Pacifico
Management Supplements
Columbia, Maryland

My only instructions from the Book Committee were: "Tell it like it was; tell it like it is; tell it like it is going to be." That assignment is best left to future historians, for it is difficult for an active participant in a dynamic function to be completely objective about it, particularly if he has a fierce affection for it. Therefore, based on my own experiences and those of some colleagues, I will try to tell it like I think it was, like I think it is, and as I hope it will be.

In addition, commercial chemical development was, is, and will be different in different companies. It is not a "thing," but a concept, and a concept flexible enough to be modified to suit the individual needs of each company's philosophy of doing business. Nevertheless, there has been a mainstream, the course of which is clear enough to follow, though it must be remembered that there were many side streams, even many eddies, during its formative period.

Like I Think It Was

From Frail Beginnings

If the nostalgia is stripped away, it is somewhat embarrassing from the vantage of our present sophistication to look back on commercial chemical development as it was in the 1940s. It now seems rather like an octopus, arms going out in all directions and not much body. It was really a collection of all those individual auxilliary functions that contributed to the sale of products, with the specific exception of statistical market research.

To put this into proper perspective, however, it must be recognized that at that time the chemical industry itself was somewhat unsophisticated and parochial. The country was at war, and so the chemical industry's orientation was toward maximizing production. Commercial problems were minimal. Even for a number of years after the war, there was a "seller's market" during which the emphasis was on filling pent-up demand, not on developing new commercial techniques. In the universities, chemistry and chemical engineering students were oriented toward technical functions upon graduation; it is doubtful that many daydreamed about glorious careers in the commercial aspects of the industry.

Nevertheless, from this most unlikely nucleus, commercial chemical development was formed. It started as a small step away from the strictly technical functions. When a problem arose in a customer's plant involving the use of a product, it became customary to send a technical expert to help solve it. The function was primarily technical in nature, and the supplier was providing a "technical service." The small step was that this function did involve the individual in contact with other companies and so introduced the need for qualities other than the technical ability—primarily those of personality. There was then, is now, and probably always will be some trepediation among sales executives at letting technical experts visit customers for fear that "some clown is going to say the wrong thing and wreck everything we built up with them over many years." Fortunately, there were not many clowns among technical service people, and even if one did pull a gaffe once in a while, it was obvious that their contribution far outweighed any disadvantages. Technical service became a distinct function.

In 1943, the Technical Service Association was formed, which became the forerunner of the Commercial Chemical Development Association (the word "chemical" was dropped in 1970). Technical service thus was the key link in the evolution of commercial chemical devleopment, and remains today one of the activities that qualifies a person for membership in CDA.

Market Development

One of the principal functions of early commercial development was that of market development. The research function at that time was the darling of the industry; they turned out new products at a prodigious rate. Many of these products were quite different from the current line of the company, so that it was inefficient to have the sales department handle them, particularly because it detracted from their primary function. Thus a new function crystallized—that of finding markets for new products. This activity required a combination of technical training and commercial skills. This balance of abilities was so different from that required in other activities that market development became recognized as a distinct function.

Developing markets for new chemicals is difficult, as shown by the fact that only about 10,000 chemicals among the millions of possibilities have sales in excess of $1,000. New techniques had to be developed to increase the probability of success of introducing a new product. The procedures used for established products had to be modified. Gradually, new procedures for obtaining sales of these new products evolved.

Market development had many successes, some of them spectacular, but it became increasingly obvious that this was an expensive procedure that could not be justified for every product that could be turned out by research. A vacuum existed, which commercial development people were fortunately well qualified to fill. Rather than simply attempting to develop markets for any new product, commercial development began to participate in the managerial decisions as to which products to market. It must be emphasized that this was not limited to statistical market research but involved all those factors that enter into a management decision of this type, including specifications desired by the market, maximum price the market would tolerate, probability of success under various conditions, and so on. Indeed, in many companies, a commercial development group was established as part of the research department, with the primary purpose of assisting research management in project selection.

Applications Research

Applications research as a separate function then began to evolve. In a product-oriented industry, it seemed axiomatic that it was the responsibility of the purchaser to evaluate the performance properties of the products offered by various suppliers. The suppliers' interests seemed to be limited to assuring the "quality"; that is, the chemical composition of the product. It seemed obviously inefficient for all potential suppliers to be gathering sales data through making the same comparisons among competitive products. Also, the laboratory data that could be obtained sometimes did not necessarily correlate with plant performance. To many it seemed a great waste of time and money.

Nevertheless, it soon became apparent that in the sale of performance chemicals, applications research data provided a competitive edge. Of course, the buyers had to redo many of the tests, but the data increased the probability the product would work for them and so in effect reduced their cost by eliminating obviously unsuitable products. Still, applications research by a supplier was, and is, very expensive. It is impractical to test most products for every possible application. Even in the areas selected, it is necessary to know which tests, if any, are significant and which conditions should be used. These questions required field contact for the answers, and so applications research was recognized as a phase of commercial development. In many companies, the applications research laboratory reported to the commercial development department.

The Commercial Development Association

The Technical Service Association started with fewer than 20 charter members. The early meetings were primarily informal exchanges of information among people in a new function who were not sure what directions their work would be taking.

The early meetings were 1-day affairs, with the participants telling each other how to conduct various aspects of commercial development. Later, programs were devoted to the chemical requirements of specific industries, and these became prime source material for market studies.

More recently, programs on broader subjects such as "Profit Trends" have become popular and more frequent.

From the beginning the meetings were informal, with a substantial portion of the time provided for personal contact among members. The members needed each other for the efficient conduct of their jobs, because an answer to a particular problem could be obtained either through one key telephone call or through weeks of effort. The esprit among members built upon the recognition of this common need exists to this day.

The operation of the association is interesting as it has been a significant factor in the acceptance and growth of commercial development. When the association was formed, relatively little power was delegated by the members to the officers; almost all important questions had to be referred back to the members for decisions. And the members did participate. They are people who have opinions on almost every issue, and the annual business meetings became something to look forward to. Never to be forgotten is the meeting where one venerable member demanded to know why the president's reception for members (a euphemism for a free cocktail party) had been eliminated that year. Told the reason was a tight budget, he promptly proposed that it be held anyway, and, budget be damned, the members voted to hold it.

As the association grew in size, additional operating authority was ceded to the Executive Committee, though all questions affecting the basic nature of the association remain the province of the general membership. The Executive Committee consists of five officers (including the past president) and four directors. Because each has an equal vote, the Executive Committee is a form of legislature; the officers carry out the resolutions of the Executive Committee, with a limited opportunity to introduce broad programs of their own. In the early 1970s, the work load on the officers and the committee chairpeople apparently became excessive. As a result, a paid, part-time executive director was employed to help with the chores, though all responsibility and authority remains with the Executive Committee.

Like I Think It Is

The Chemical Industry

Commercial chemical development is not, of course, an independent entity complete unto itself; it must adjust to the industry it serves. Thus to understand the stresses and strains within commercial development as it evolved to satisfy the changing needs of the chemical industry, it is first necessary to examine the principal trends within that industry. In the early days, the industry was primarily concerned with products made by classical chemical synthesis. When these chemicals were in turn used for their physical properties—for example, as paints, pharmaceuticals, cleaning compounds, and so on—those operations were considered related, but distinct industries. There has been indeed a general grouping known as the "chemical processing industries," but this by no means includes all the operations that are primarily chemical in nature.

More recently many chemical companies extended their activities into applications where the properties, as opposed to the chemical reactions, of their products were used. Outstanding examples are synthetic fibers and films, but there are innumerable less spectacular projects. In some cases, being unskilled themselves in the marketing of such products, the chemical companies purchased companies in other industries. Sometimes they developed the necessary strengths internally. Through this combination of internal growth and acquisition, the chemical companies gradually pushed into other industries. In turn, other companies found that their basic products were really chemicals (for example, petroleum and starch), and they began some chemical synthesis operations. Thus the traditional boundaries between industries began to crumble, to such an extent that some chemical companies found they were in so many other businesses that they removed the word "chemical" from their names.

A more subtle trend in the industry that affects commercial development has been the increasing minimum economic size of plants for making many commodity chemicals. This increases the importance of proper business planning, including economic forecasting many years into the future.

Performance chemicals have also been affected, but in a different way. While attention has been focused on the population explosion, there has been occurring, almost unnoticed, an information explosion. One important effect is that it has made replacement of performance chemicals less risky, thus resulting in a shortening of the life cycles of this type of product. Although this makes market development of new products somewhat easier, it makes overall business planning more difficult, because recapturing the substantial investment involved in the development of products with short lives is a formidable problem.

Also, research has become more expensive in recent years, and some feel that it is less productive.

Many companies have therefore shifted emphasis somewhat from research of new products to process improvements for the giant plants they already have, where even minor improvements in cost per pound of material can generate more profit at less risk than new products can. The need for commercial development activities is diminished proportionately.

Commercial Development Departments

In many companies, it gradually became apparent that the auxiliary aspects of commercializing a product were scattered over many departments. In some companies, therefore, commercial development attained a broader, more unified status. A separate department was established that included responsibility for participation in selecting suitable projects, introduction of the new product, and market development until commercial sales were established. This involved the functions of market research to gather data, market development to promote the products, and often support laboratories in applications research or technical service. Some companies included the function of "industrial intelligence," whose responsibility it was to keep aware, from published data, of what the industry was doing.

Some companies are still at this stage of the evolutionary process. Others, however, have progressed further. Rather than limiting the planning of the department to how to handle specific projects, the responsibility was broadened to include all of corporate or divisional planning. Sometimes this function was incorporated into the commercial development function and sometimes it was separate, but the function itself quickly became recognized as a province of commercial development and became accepted as a type of experience that qualified the individual for membership in CDA.

From the broad responsibility of business planning, it was only a short step to responsibility for implementing the plan. One of the ways in which new business activities can be achieved is by means of mergers and acquisitions, and so this function also became an integral part of commercial development.

Of course, there were always acquisitions and mergers; some of the largest chemical companies were formed by mergers before World War II. Still, it may surprise younger people to know that mergers received relatively little thought in the late 1940s and early 1950s. There was a strange philosophy then that mergers should be undertaken only when they advanced the company toward its business objectives and so increased the return on investment. Then someone discovered that even mergers that did not improve return on investment might improve earnings per share, and this increase in turn gains a higher p/e multiple, which permits purchase of still other companies that have lower p/e ratios. A money machine apparently had been found; it was some years later before it became apparent that the procedure is related in principle to chain letters.

In any case, one major aspect of acquiring another company is an appraisal of its future business prospects. Aside from top operating executives, probably no one in the organization is better qualified for an overall examination of a business than persons trained in commercial development. In some companies the function eventually became part of the commercial chemical development concept. In other cases, commercial development persons became executives in charge of that function, which remained separate from commercial development in the company's organizational structure.

Today commercial development usually is concerned with all aspects of new business activity, and some companies are adopting names for it like "new ventures" or "new activities" department, in many cases with corporate vice-president stature.

Status of Original Functions

As commercial development adopted these areas of broader responsibility, the emphasis on most of the original functions began to decline. None of the functions was abandoned, of course, because they remain essential to the operations of the company. Rather they tended to crystallize into subfunctions, handled by junior personnel in the commercial development department.

Although market research will always be an important aspect of commercial development, there are forces pulling it in other directions. Statistical market research is also of great value in activities on present products, a function essentially separate from commercial development. Also, it has become recognized that there are many more aspects to the sale of products than simple statistics and that data on these aspects can be efficiently gathered by the same people doing statistical market research. This function, therefore, has been considerably broadened to include such factors as channels of distribution, discount schedules, warehouse locations, and so on. To emphasize this change, the name of the function has been changed to "marketing research."

Another factor that tends to set this function somewhat apart from commercial development is that it has its own association, The Chemical Marketing Research Association. There is some overlap in membership with CDA, and there have been frequent discussions in the past on the merits of merging the two groups. However, each has set its course in a somewhat different direction, and although cooperation between them will undoubtedly continue, it now seems likely that no merger will take place in the foreseeable future.

Market development continues to be an integral part of commercial development. It has benefited from the many expensive lessons of the past and is conducted with considerably more skill. It is now relatively rare for a company to make products randomly and simply offer them to the industry as "products with undeveloped commercial potential." Indeed, that procedure is restricted almost entirely to derivatives of some raw material a company would like to use up. Today the probable success of new products is carefully analyzed in advance. Unlike the shotgun approach of the past, these new products are rifled at specific targets. Market development is now recognized throughout the industry as an expensive procedure with complete loss of investment if the product fails; it is used only when the substantial investment required appears worthwhile.

In some companies applications research has undergone a most interesting transition in recent years. It still performs its original functions of developing data to improve the chances of success in selling current products, in developing data to assist in the sale of new products, and in developing improvements in these products. However, it has added a new dimension. In many companies the applications research personnel become experts in the needs of the industries they serve. They get to know a great deal about the trends developing in those industries. Thus they often are able to suggest to their companies the types of new molecules that should be developed to suit these future needs.

For example, a company may be selling a fungicide for a paint. Although paint will remain a product of commerce indefinitely into the future, the chemical composition of paint changes with some frequency. The applications research department can detect that the trend away from one type of latex paint, in which their present fungicide is compatible, to a new type, in which their product will not be suitable. This gives the supplier time to develop a new product that will be useful in the new system and so prevents loss of a market.

Technical service, however, is now seldom considered an integral part of commercial development, and few companies have this function in that department. Indeed, technical service has so declined in importance in commercial development that it would no longer qualify a person for membership in CDA, except for deference to the origins of CDA.

International

It may be difficult to imagine, but there was a time when a business trip to Europe was worthy of note. Following the stripping of German technology after the war, there was a period of several years of relative inactivity in international affairs. As the shattered economies of Europe and Japan were rebuilt, contacts increased rapidly. Licensing of U.S. technology to these countries could be handled by sales people, but the reverse procedure often involved commercial development to evalutate the potential of the products offered. Still later, as U.S. companies increased the number of plants and joint ventures they installed in other countries, it was found that the contacts and knowledge gained by commercial development people in the licensing procedure was again useful. Indeed, several commercial development people now have responsibility for managing plants in other countries, and it is impressive to note how many top executives in the international departments of U.S. companies have commercial development backgrounds.

Relations with Management

One of the most interesting aspects to reflect upon has been the relationship between commercial development and top management. In the early days the relationship was simplicity itself; management hardly knew that commercial development existed in the company. The prestige of commercial development grew partly through its accomplishments but also in part because of the advancement of individuals with commercial development backgrounds. Convinced of its merits, they carried the function up the corporate ladder with them.

For a brief span of time, it seemed that the proper function of commercial development had been recognized. That is, growth of the corporation is one of the primary duties of the corporate president. Because in companies of any size this is beyond the abilities of any one person, the responsibility for planning, gathering data, and making recommendations for growth, as well as the first stages of implementation must be delegated. The commercial development department seemed to be the logical candidate for this responsibility, becoming, in effect, an extension of the president.

This arrangement has been set up in several companies and appears to be working satisfactorily. In other companies, however, the prestige of commercial development has suffered a noticeable decline. Both management and commercial development personnel speak of communications problems, which is another way of saying that they don't understand each other. This is incredible, of course, when commercial development is understood to be only an extension of the president, yet in some companies commercial development is so far down the line that it does not even have direct access to the president, the person it is supposed to be part of.

Part of the problem, however, lies in the general disenchantment of management with staff functions as the profit squeeze intensifies. They are less eager to make the substantial, relatively risky investment into new products when they could be saving a little insurance for the less risky acquisition route. Commercial development departments that have recognized this trend and evolved into groups geared to new ventures have prospered, while those clinging to old virtues and old concepts are having rather tough going.

The Commercial Development Association

As the boundaries of the chemical industry crumpled, two effects developed. Many companies that were primarily chemical in nature became major suppliers of other products. At about the same time, companies in other technical industries found they were encountering the same problems of product development that the chemical industry had encountered and solved a generation earlier. Because the functions of commercial development are essentially the same whatever the technical product, some of them employed commercial development executives from the chemical industry. Thus members of the association now are employed by metals companies, pharmaceutical producers, and even investment bankers.

Recognition of this change implied that the association had an importance beyond that of the chemical industry, and so dropped "chemical" from its name in 1970. However, application for membership for individuals in other industries has been slow, so that despite the change, the membership of the association still consists primarily of individuals with a chemical background.

Like It May Be

What does the future hold for commercial chemical development? This will depend primarily on the directions taken by the chemical industry.

Easiest to predict are are extrapolations of current trends. Certainly the boundaries between the chemical industry and the industries it serves will continue to fade away. It seems likely that eventually most companies now considered chemical companies will range in scope from extraction of the raw material from the earth through all the processing and marketing steps until the material is finally used up by some type of consumer.

This trend will obviously have a fundamental effect on the nature of commercial chemical development. Emphasis on the strictly chemical aspects will decrease. Fortunately, this is not in itself a problem because the techniques of commercial development will be just as appropriate in other technically oriented industries. However, this development will require a broader knowledge of the fundamentals of other industries and a wider acquaintance with executives in those industries if the commercial development person is to perform his or her job adequately.

Because many companies will be their own best customers for intermediate products, the importance of industrial market development will decline. On the other hand, market research and market development of consuming markets will correspondingly increase in importance.

The increasing size of the companies will itself become a factor. As companies become larger, they become less maneuverable. Planning will thus increase in importance. Indeed, improved techniques will be required because very large companies will have to plan many years in the future, and in much more detail, because they will not be able to shift course easily if the circumstances that develop are not those that are anticipated.

Techniques will have to be developed to test the future. One that appeals to me is that of sending out advance parties to scout areas of new technology. For example, this might take the form of major companies supporting inventors or small companies in new areas of technology. Based on the success of these outposts, the company can decide whether to pursue that area or not. Indeed, if it has several such scouting parties, it can select the more promising areas for major investments. Commercial development will be needed both to find and to evaluate these new fields of activity, and perhaps to help manage the new company in its formative stages.

In any case, commercial development has not yet developed the fatal flaw of excessive specialization but remains flexible enough to evolve into any role required of it as the industry continues its fascinating course of unending change.

2 ORGANIZING FOR AND MANAGING COMMERCIAL DEVELOPMENT

James W. Leenhouts
Michigan Division
Dow Chemical U.S.A.
Midland, Michigan

Samuel W. Tinsley
Union Carbide Corporation
New York, New York

There are two aspects to managing commercial chemical development. One aspect is concerned with the managing of the commercial development function from an administrative viewpoint, whereas the other is concerned with it from a philosophical and process aspect. During the 1950s and early 1960s, the commercial development function was usually performed within the aegis of a formalized commercial development department.

Within the past few years, however, commercial development departments as such have ceased to exist in all but a few companies. In a survey taken in 1972, it was found that most of the members of the Commercial Development Association, although carrying out many of the activities commonly associated with commercial development, now perform their activities outside the framework of a formalized structure or work within the framework of other departments. Nevertheless, the function itself, as a process, has not changed.

The necessity, then, for the efficient management of commercial development has become even more important today than was formerly the case. This becomes especially true in view of the complex nature of changing technology and the shortened time frame from research to commercial reality in today's competitive world. Given the high rate of product introduction failure and the need to manage the generation, evaluation, and development of a stream of ideas for new products, it is not difficult to understand the crucial nature of these decisions affecting the management of commercial development.

One of the problems concerning the organization of the commercial development function lies in the nature of the corporate organization itself, which is essentially geared to the company's current product lines, so that emphasis is on clear-cut lines of authority and responsibility. On the other hand, commercial development combines qualities that are foreign to those found in the rest of the company organization, such as the need to establish a flow of ideas, their evaluation, and their development. The process is at once continuous and discontinuous. Also, the qualities found in the commercial development practitioner are different from those of the line manager. It follows, then, that the commercial development function would be better located outside the rest of the corporate structure. As was pointed out in the first edition of this book, many companies in the past have centered this function in an autonomous group.

This group may be a section, division, or department, and generally reports directly to the chief executive. It always has direct responsibility for finding and evaluating the profit possibilities of new products. A growing trend is to charge it with full responsibility for organizing and coordinating the entire program of developing, evaluating, and finally commercializing new products. The companies that use this system have been notably successful in commercializing new, profitable products at a constantly increasing rate.

The degree to which an ideal commercial development organization can be achieved in a specific company depends upon many factors. It must be recognized that organization alone cannot guarantee success. The capabilities of key individuals in the major departments of the company and their ability and willingness to work as a team with the commercial development group are the dominant factors that determine success or failure. Organizational structure and lines of responsibility should and must be changed, depending upon the capabilities and limitations of the individuals. The departments to which various phases of the development work are delegated will vary according to the characteristics of the individuals with which they are staffed. Commercial development is not static. Theories and practice change. An alert organization keeps abreast of the changes. It will probably never be possible to resolve completely the question of the best method for solving a specific problem or finding ideal individuals to fit an ideal organization. The most important point is to recognize important functions and properly delegate authority to a group that will see that decisions and policies are aggressively and intelligently carried out.

Where the commercial development function finally resides depends on management philosophy, how large a product development program the company needs and can afford, the complexity of these projects, and the current organization structure.

The commercial development activity can be located at various organizational levels as follows: (1) corporate level; (2) divisional; (3) corporate and divisional. By function, the activity can be located within: (1) marketing; (2) research and development; (3) a separate new-products unit.

Recently, the Conference Board conducted a study of 108 companies, of which 68 were manufacturers of industrial products and 40 were producers of consumer products. The results were reported in *Options in New Product Organization.** The advantages and disadvantages of locating the commercial development activity by organization level and by function are summarized in this section.

Location by Organization Level

Location at Corporate Level

Advantages

1. Allows better coordination and control of all the company's programs for innovation.
2. Resources available for new-product development can be concentrated where the outcome is likely to be most fruitful.
3. New-product decisions are less likely to be colored by the considerations of divisional management.
4. New-product ideas need not necessarily be tied to the kinds of products, technologies, and markets with which individual divisions are familiar.
5. Senior corporate management is kept more aware of current new-product activities, with the result that these will probably receive greater attention and support.

Location at Divisional Level

Advantages

1. New products developed at the divisional level are more likely to reflect more realistically customer needs, market opportunities, and productive capabilities of each division.
2. Division-wide involvement and commitment to the new-products program may increase the chances of success.

Location at Both Corporate and Divisional Level

Advantages

1. Offers flexibility in allocating company-wide resources for new product development between corporate and operating levels.

*Conference Board Report 613, 1974, by David S. Hopkins.

2. Policy control and perspective can be retained at the corporate level, while the divisions can continue to play an active role in new products programs.
3. Helps to engender an innovative environment throughout the company.

Disadvantages

1. Responsibility for new products is divided, perhaps causing problems in communication, duplication of effort, or missed opportunities.
2. Difficulties may arise in making a logical allocation of responsibilities between the two levels.
3. Unless there is an effective control system, the number of products under development may turn out to be too many or too few.
4. This arrangement is suitable only for firms of substantial size that are well staffed at both corporate and divisional levels.

Location by Function

Location Within Marketing

Advantages

1. The marketing department is close to the marketplace and sensitive to market opportunities, and so products are likely to meet customer needs.
2. Planning and control of new product activities are centralized in competent hands that can carry products all the way to market.
3. Development of new products is likely to be carried out with urgency and drive.

Disadvantages

1. Marketing personnel may lack broad, balanced perspective, especially with respect to technical and other nonmarketing considerations.
2. There may be excessive emphasis on "me-too" products or on only modest improvements in products for sale only to familiar markets.
3. New-product efforts may tend to be short range and hurried, with objectivity perhaps clouded by enthusiasm.
4. If regular product managers are assigned responsibility for new products, in addition to existing products, the new products program may not be given enough attention or it may be handled in an incompetent manner.

Location Within R&D or Other Technical Unit

Advantages

1. New products are more likely to be innovative in a technical sense.

2. Creativity and productivity of R&D personnel may be enhanced.
3. Technical feasibility of manufacturing a new product is perhaps more likely to be established at an early stage.

Disadvantages

1. R&D may not be sufficiently aware of market needs and other nontechnical considerations.
2. There may be a tendency to spend time on the development of "technically fascinating" products, with less regard for practicality or company objectives.
3. New product efforts may tend to be protracted and carried out in a perfectionist manner.

Location Within a Separate New Products Unit

Location within a separate new products unit is another alternative that is being used in a number of companies.

Of 57 divisionalized industrial products manufacturers surveyed by the Conference Board, a new products unit was reported at the following levels: corporate level, 26 percent; group level, 11 percent; and divisional level, 14 percent. Of 11 nondivisionalized companies surveyed, 9 percent had separate new products units.

Most of the companies surveyed felt that this type of organization is useful only when a company has the resources necessary to sustain a large and continuing new products program, and under the following circumstances:

1. When new product projects are costly and complex and will extend over a long period of time.
2. When there is a high element of risk.
3. When the new product projects are unrelated to the existing product line or will move the company into unfamiliar markets or technical areas.

The advantages and disadvantages of this type of organization can be summarized as follows:

Advantages

1. Responsibility for new products is clear-cut and is not diluted by the current activities of the company.
2. The unit may be more objective.
3. The new products effort will have higher visibility than under other types of organization.
4. Emphasis is not on short-term results.

Disadvantages

1. The unit runs the risk of becoming isolated or pursuing impractical projects, by virtue of its being a separate unit.
2. The possibility exists for friction between the group and the functional departments.
3. Because of its high visibility, the group may be susceptible to staff cuts in time of recession.
4. Recruiting the specialized leadership and staff may be difficult.
5. The problem of interfaces between the new-product group and other functional groups may lead to loss of time and momentum.

Generally speaking, a separate new-products group seems best suited to the development of high-technology industrial products.

Within the above framework of location of the new product function, the question arises as to how project teams and task forces operate. Generally speaking, when a company turns over the main responsibility for new product coordination to a separate new products unit, or to units at either the corporate or divisional levels, the intent is that these groups should initiate, or be responsible for, a continuing program of development. Sometimes, however, a project is encountered that is too complex for it to be handled under the normal aegis of the new products group. On these occasions an ad hoc task force will be set up to handle the specific project. The task force is usually placed outside the new product department's organization structure and contains specialists from various functions from elsewhere in the company who possess the requisite skills needed to ensure the successful completion of the project. These task forces have been called variously "project teams" or "new business venture groups," and will be described shortly.

The advantages of the task-force approach lie in its flexibility. The company can utilize the group to accomplish major projects that would ordinarily tie up the resources of the new products department for an extended period, such as enabling the company to enter a new business area. Another major advantage lies in the ability to gather the company's ablest personnel from different functional areas to approach a problem with a fresh approach. A certain synergism can be attained with this use of teams, and the chances that a project will be successful can be enhanced. Also, by keeping the task force outside the organizational structure of the company, the red tape often encountered in large companies can be circumvented and restraints on creativity removed.

On the other side of the coin, there are a number of disadvantages inherent in this approach. There is first the problem of finding talented individuals from within the company and then persuading them to join the task force, which may be perceived by the individual as being peripheral to his or her main career path. The department heads giving up the personnel may also be reluctant to do so.

Another not so obvious shortcoming lies in the independence of the task force. Here the problem is to allow the group sufficient independence to accomplish its task, while meeting the desire of top management to keep the group under control. There is also the possibility that the task force may become so committed to the project that it loses its objectivity. Finally, because of the group's high visibility and the length of time involved in completing a complex project, it may very well be the first object of cost-cutting campaigns, which occur with a certain regularity.

The new business venture group is usually considered to be somewhat similar to, but an extension of, the project team. The dissimilarities are found in the purpose and the size of the venture group. The project team is usually a loosely structured task force assembled for a limited purpose, whereas a venture group is more likely to be tightly structured, with its purpose broad enough to ensure its autonomy and its support by top management both in terms of money and prestige. It is more likely to be given the task of getting the company into a new business area.

New Product Committees

The role of new product committees has varied widely over the years. Their function was explored thoroughly in the earlier edition of this book, and the viewpoint expressed then contained the same reservations about their usefulness as does the Conference Board survey referred to earlier in this chapter.

There is considerable disagreement as to the usefulness of these committees. Like all committees, these committees are subject to all the shortcomings attributed to them such as their characteristic inability to make decisions. Those in favor of committees feel that new product proposals should be evaluated by a separate, detached corporate staff group. For example, the R&D director of a chemical company surveyed by the Conference Board believes that appraisal of all new product proposals is best performed by "a separate, detached corporate staff group of economists and marketing strategists." (In his compnay, such a corporate group is assigned responsibility for measuring the return on investment and growth prospects of each new product proposed at either corporate or divisional levels, in comparison with all other new product or investment opportunities available to the company.)

In some companies new product committees operate at the top-management rather than the middle-management level, with the committee being made up of the heads of key functional departments. It has also been suggested that the new product type of structure would be useful in a small company that would not be able to afford a new products group.

It is generally agreed that a new product committee, if it is to operate effectively, should have a clearly expressed charter of its responsibilities. It should work with a formal agenda and keep minutes of its meetings. Its chairperson should be strong, preferably a senior executive or the head of the department having new-product planning and development responsibility. The committee-membership should include representatives of all departments having significant involvement in the new-product programs.

Some perceive the committee's role as a brainstorming group creating new product ideas, whereas others look at such a group as being a review board for new product proposals. Most companies having such committees use them for broad purposes such as communications, coordination, or for decision making at various stages in new-product development. In terms of decision making, the major criticisms of these committees are that they are slow in making decisions and that, because of the penchant of companies for compromise, they generally make weak ones. However, because product development is necessarily an individual type of effort that is inherently fraught with risk, committees can be useful. If the committee is composed of objective individuals, all possible ramifications of a project can be explored, and objective evaluations reached.

The tables summarizing the Conference Board study referred to in the preceding section are shown at the end of the chapter. Of the 108 companies, 68 are manufacturers of industrial products and 40 are manufacturers of consumer products.

The Conference Board study notes that neither the aggregate group of 108 companies nor each of the two major subgroups constitutes a random sample of such manufacturers in the United States. There is evidence, for example, that the survey generated an above-average response from executives in larger-sized companies. Furthermore, the relative dependence of the companies on new products for current sales volume is generally higher than that found in earlier Conference Board surveys among groups of companies known to represent a reasonable cross section of all but the smallest firms in the manufacturing sector. One explanation could be that, in general, companies are becoming more dependent on new products than before. Although this may be true, it seems likely that a substantial number of the companies to which the table data refer are among those placing exceptional emphasis on new product development.

In order to illustrate the key points of the preceding discussion, two case studies of how commercial development has changed over the past two decades at two major companies, Union Carbide Corporation and Dow Chemical Co. are presented on the following pages. Other examples will be found in Chapter 5, "New Venture Management," and in Chapter 12, "What Top Management Expects from Commercial Development."

Union Carbide Corporation

When Hoyt Corley's classic *Successful Commercial Chemical Development* came off the press in 1954, Carbide and Carbon Chemicals Company was pursuing a development approach that was a refinement in degree, not in kind, of the methods instituted at the very beginning of the company in the early 1920s. Indeed, it was 1924 when the first chemist salesman in the industry, J. G. Davidson, was appointed Manager of Chemical Sales. His policies included training chemists and chemical engineers not only to be salespeople but also to recognize potential markets and to teach the company's customers to use its products correctly. These technical representatives were backed up by R&D and technical services functions that exhibited a remarkable ability to synthesize molecules on demand and by engineering and production functions that somehow turned the molecules into commercial products. In this manner the aliphatic petrochemical industry came into being.

To give some perspective to this developmental approach, during the 35 years from 1925 to 1960, the company sold at least 388 chemical products, of which 210 were first produced in the United States by Carbide. The success of the approach was attested to by the growth of the company and, significantly, by the fact that the first CCDA Honor Award went to Harry McClure of Carbide and Carbon Chemical Company in 1950.

A number of factors began to come into play during the late 1950s that rendered this commercial development method (know your customer, determine his or her problem, and solve it with a new product) inoperable. These included: the volume growth of chemicals and competition, which forced manufacture out of multipurpose small-to-medium-sized equipment into dedicated and designed units with higher investments but lower operating costs; the resultant greatly escalated penalty for being wrong about a new product; the increasing costs of R&D, particularly process development; the emergence of safety in process design and product toxicology as significant factors in new product decisions; and the need for the ability to plan in an orderly manner the growth of the business in order that *all* of the ever-increasing resource requirements might be deployed in an optimum way.

Therefore, in 1962, the Chemical Company was reorganized in order to move toward a market-oriented approach to both present business extension and expansion as well as new business development. Coincidentally, there followed 3 years of sustained growth in sales and earnings in the industry. Thus, it was with a high degree of confidence that both the organization and the market-oriented approach were adopted by the now-combined Chemicals and Plastics Group in 1967—just in time to experience the infamous "blood bath of the 1960s" in the industry.

Forging in the fires of overcapacity and declining prices and the resultant disastrous earnings, though anything but pleasant at the time, has led to the Carbide of today. The company, for the first time on a rigorous basis, has decided what it is, where it wants to go, and how it is going to get there. All present businesses have been characterized and categorized into Strategic Planning Units for purposes of resource allocation and management objective; a uniform management system has been institutionalized; long-term strategies have been enunciated and tactical management has been elevated to the Corporate Senior Vice-President level. All diversifications, which include new products and new businesses, are the responsibility of a Corporate Executive Vice-President, who sits on the Management Committee, whose prime responsibility is strategic planning.

The new business approach is market oriented, usually with a requirement for a high technology input. Emphasis is placed on defining and meeting needs in market niches, market discontinuity, or in markets where the *real* need is perceived to be different than that met by the competition. Emphasis is also placed on defining needs in terms of total functions and fulfilling those needs with total systems. Most importantly, instead of tolerating a commercial development activity until it gets highly visible in size or resource requirements before making a judgment on the appropriateness of the activity for Carbide, the Management Committee endorses or rejects new businesses at an early stage, lives through the ups and downs with the commercial developers, and commits the resources to expand and develop the business as an integral part of the strategic plans of the Corporation.

The Dow Chemical Company

One cannot deny that Dow has prospered, almost miraculously, over the last 20 years. Growth of 14 percent compounded in the chemical business obviously indicates that commercial development has been successful. The changes over the 20 years, then, are worth examination. Two major factors have been at work causing change: (1) changes in the type of effort; and (2) changes in the enabling skills. Each will be examined separately, but the reader is cautioned to observe one very important point. The process of change is continuous. It did not start in 1958, nor did in end in 1978. It is as active today as it was in 1956. Some areas of the diversified business of Dow are changing much more rapidly than others and are in the forefront of change. Therefore, a composite picture is about the best one can hope to do.

Changes in the Type of Effort

Twenty years ago, the commercial development effort was individual, diversified, and entrepreneural; today it is a team effort that is well integrated and planned. Dow has never had a formal commercial development department. The closest internal organization to this is the Technical Service and Development Departments associated with the major businesses. However, while these Departments, called "TS&D," were responsible for the development of many Dow products and businesses, they were originally a laboratory-oriented group designed to provide technical service to the marketing function. Containing the same individual spirit that pervaded the whole Dow organization, members of TS&D were in a unique position to see a customer's need on one hand and visualize the technical solution on the other. This was the organization in the middle 1950s.

The underlying major change after the middle 1950s was the evolution in Dow of the "business concept." Originally developed in the creation of the Dow Plastics Department in the early 1950s, the business concept developed to encompass all of the Dow businesses by the late 1950s. The business concept was built on a series of "teams" with the following hierarchy, with each reporting to the one above (with example):

Dow management
↑
Department management team (Chemicals Department)
↑
Business management team (oxides and derivatives)
↑
Product management team (alkanolamines)
↑
Task force (alkanolamine quality)

Each team had four members, one each from research, marketing, TS&D, and production. Each team met about once every other month to establish the appropriate strategy and resources needed to cause the product or business to grow. Profits, manpower, projects, processes, and competitive assessments were all made. Since line managers in each of the functions were members of the team, action could be quickly undertaken. Obviously, the teams were a very powerful growth stimulant because the functions were communicating and making decisions all at the appropriate level.

The team was considerably more effective than the individual effort it replaced for another reason—project size. Prior to this time, only a few individuals, largely through their personality, were able to penetrate the multifunctional maze and create a new internal enterprise. Such individuals were enormously successful and created the base for the present Dow Chemical Company. Nevertheless, the team, on the other hand, could mount a larger effort and had greater resources available. Project size could be increased. Thus, the strong personality of the 1950s has now been replaced by the "team" person. Efficiency has increased, but, as a penalty, the creation of diverse and entrepreneurial business do not occur with the same rapidity.

Changes in the Enabling Skills

Product testing in the 1950s was really only fully practiced in a few areas. Today, a Dow laboratory can probably be found to evaluate a possible contender for almost any use. Now able to evaluate the use of most of its products in its own laboratories, Dow, and the entire chemical business, has become highly skilled at *their customer's business.* Because profit is likely to fall with the innovator, the chemical business has been able to grow by creating growth in the industries it serves. For example, the success of synthetic fibers is not a textile industry development any more than the success of fiberglass boats is a development of the boat yards. Much success in the ability of the chemical business to learn their customer's business is the rapid development of *standardized tests.* As a result, a common "language" or "specification" has been developed. Fold, rupture, wrinkle, impact, whiteness, UV resistance, flame spread, tensile strength, and wet strength are just a few of the broad test classifications available now to the developer who seeks to grow in an area.

Accounting and financial data at Dow is now available in detail on each product from plant to plant on a monthly basis before the fifteenth working day of the next month. For example, thanks to computers, even the number of research hours on any one single problem number is computed into the Monthly Product Income Statement for a product. Furthermore, the data are compared with a detailed plan established at the beginning of the year. The revolution in accounting and finance has made accurate planning a real tool. With this, growth is a natural outcome. Valuable resources are not wasted on products that are declining in health; conversely, where resources can compete to reverse a downward trend, new projects are initiated early.

Increases in *analytical and toxicological* skills have probably led other technical skills. At Dow, the ability to assess product quality and product health effects has allowed the company many very profitable opportunities to enter a business where the customer or the competitor could not offer such assurances. Likewise, these skills have led to the product stewardship program where the company assesses the downstream impact of a product and may refuse to sell a product if it is not satisfied it will be used properly. Obviously, even a minor brush with health or environmental problems, real or otherwise, is sufficient to cause a promising development to go down.

Lastly, *government regulation,* seeming the great scourge to growth, has provided a new basis for growth. Fire retardants, high-performance brake fluids, furniture cushioning, and pollution control chemicals, are a few of the many products that owe much of their successes to government regulation.

In summary, a great deal has happened in 20 years to improve commercial development. Perhaps the only thing that has remained constant is the attitude of Dow management, that *growth is the only acceptable solution to corporate health.*

TABLE 1

Location of New Product Activities in 87 Divisionalized Companies, by Organizational Level and Function

Level and Function	Total Divisionalized Companies (87)	Manufacturers Industrial Products (57)	Companies Heavily Dependent on New Products (62)
Percentage of divisionalized companies having new product activities located at:			
Corporate level (total %)	71	77	73
Marketing department	20	23	19
R&D department	38	44	37
Separate new products unit	24	26	26
Other	8	7	10
Group level (total %)	47	40	48
Marketing department	21	16	26
R&D department	21	16	26
Separate new products unit	9	11	8
Other	7	5	5
Divisional level (total %)	89	89	92
Marketing department	49	49	55
Separate new products unit	15	14	15
Other	7	7	8

Source: Conference Board

TABLE 2

Location of New Product Activities in 21 Nondivisionalized Companies, by Function

Function	Total Companies (21)	Manufacturers Industrial Products (11)	Companies Heavily Dependent on New Products (10)
Percentage of nondivisionalized companies having new product activities located in:			
Marketing department	52	64	50
R&D department	47	55	40
Separate new products unit	9	9	10
Other	9	–	10

Source: Conference Board

TABLE 3

Use of Regular Staff Product Managers for Existing Products in 108 Companies

Use of Regular Staff Product Managers	Total Companies (108)	Manufacturers Industrial Products (68)	Companies Heavily Dependent on New Products (72)
Have regular staff product managers who plan for existing products	84	85	85
Do not have regular staff product managers who plan for existing products	14	13	14
Not reported	2	2	2

Source: Conference Board

TABLE 4

Responsibility of Regular Staff Product Managers for New Products in 91 Companies

Regular Staff Product Managers	Total Companies (91)	Manufacturers Industrial Products (58)	Companies Heavily Dependent on New Products (61)
Planning for existing products:			
Responsible for	100%	100%	100%
Planning for more than routine modifications of extension of their lines:			
Responsible for	68	74	77
Not responsible for	32	26	23
Planning for major new products:			
Responsible for	46	50	53
Not responsible for	49	45	46
Not reported	5	5	2

Source: Conference Board

TABLE 5

Use of Project Teams and New-Business Venture Groups in 108 Companies

Use	Total Companies (108)	Manufacturers Industrial Products (68)	Companies Heavily Dependent on New Products (72)
Use of project teams:			
Use routinely	15	16	18
Use occasionally	61	65	64
Never use	22	18	18
Not reported	2	1	–
Use of new-business venture groups:			
Have such groups	36	35	39
Do not have such groups	63	65	61
Not reported	1	–	–

Source: Conference Board

TABLE 6

Use of New-Business Venture Groups in 87 Divisionalized Companies, by Organizational Level

Use and Level	Total Companies (87)	Manufacturers Industrial Products (57)	Companies Heavily Dependent on New Products (62)
Have new-business venture group(s) (total %)	45	42	45
At corporate level	33	30	32
At group level	6	4	6
At divisional level	14	11	15
Do not have new-business venture group	55	58	55
Total	100	100	100

Source: Conference Board

TABLE 7

Use of New Products Committees in 108 Companies

Use of New Products Committees	Total Companies (108)	Manufacturers Industrial Products (68)	Companies Heavily Dependent on New Products (72)
Have new products committee(s) (total %)	60	60	60
Usual purpose is communications	32	34	36
Usual purpose is coordination	37	37	40
Usual purpose is decision-making	25	25	25
Do not have new products committee	37	37	40
Not reported	3	3	–

Source: Conference Board

TABLE 8

Use of New Product Committees in 87 Divisionalized Companies, by Organizational Level

Use and Level	Total Companies (87)	Manufacturers Industrial Products (57)	Companies Heavily Dependent on New Products (62)
Have new products committee(s) (total %)	57	60	53
At corporate level	29	30	26
At group level	14	11	16
At divisional level	43	49	40
Do not have new products committee	40	37	47
Not reported	3	3	0

Bibliography

"Options in New Product Organization." CBR 613, 1973.

The following are from a symposium conducted on "Changing Times in Commercial Development," March 9, 1971:

Blake, Edmund J., Vice-President, Pioneer Lands Corporation. "Venture Capital and Corporate Development."

Canavan, William J., Vice-President and General Manager, Olin Corp. "What Management Expects from Commercial Development."

Donley, Edward, President, Air Products and Chemicals, Inc. "Commercial Development at Air Products and Chemicals, Inc."

Fleming, Samuel C., Director, Arthur D. Little, Inc. "Trends in Commercial Development for the 70's"

Goldmark, Dr. Peter C., President and Director of Research, CBS Laboratories. "Cities of the Future."

Graziano, Frank J., President, Crompton & Knowles Corporation. "Commercial Development Abroad to Maximize Profits at Home."

Hatfield, Marshall R., General Manager, 3M Company. "When the Heat Is on—Stay Cool."

Reiser, Robert F., Manager, Xerox Corporation. "Commercial Development under Fire."

Todd, Robert W., Vice President, First National City Bank. "Return to Return."

The following are from a symposium conducted on "Planning for Profitable Growth," October 12, 1970:

Avery, Henry, and McNeil, J. William. "Strategy Planning for Growth."

Collier, James R. "Effective Long-Range Business Planning."

Throdahl, Monte C. "Corporate Objectives and Growth."

The following is from a symposium conducted on "Evaluating the Effectiveness of Commercial Development," June 1971:

Kelly, Ralph E., Director of Development, Hercules, Inc. "Evaluation of New Ventures Today."

Peterson, Russell W. "New Venture Management in a Large Company." *Harvard Business Review*, no. 6 (November–December 1967), p. 68.

3

INNOVATION AND CREATION OF NEW PRODUCT IDEAS

Newman H. Giragosian
Delphi Marketing Services, Inc.
New York, New York

New product ideas are even more important today than was the case two decades ago, when the first edition of this book appeared. As that edition stated, "Everybody singin' 'bout Heaven ain't goin' there." As was stated in the preface to this edition, industry has been undergoing radical changes during the past twenty-five years.

Forces that have been changing industry are:

1. Rising obsolescence rate of products
2. Rising operating costs for labor, materials, and plant
3. Accelerating rate of technological change, although the acceleration has slowed down considerably in recent years
4. Increasing cost of research and development
5. Federal and state government action concerning patents, taxation, tariffs, anti-trust, profit margins, new plant investment incentives, resource pollution, and public health
6. Intermittent periods of excess plant capacity
7. Declining profitability on both product and corporate levels

Origins of New Product Ideas

New product ideas are generated from a number of sources. Table 1 presents one scheme for examining them. *Internal* sources include top management, research and development, marketing research, marketing, engineering, manufacturing, and acquisition groups, planning, and new venture analysis

TABLE 1
Sources of New Product Ideas

Point at Which Idea Originates or Enters the Organization	Source of Idea
Sales force	Knowledge of customer needs and applications
	Inquiries from customers and prospects
	Familiarity with competitive, allied and substitute products
Marketing research	Field calls to determine consumer needs through discussions with:
	Research directors
	Marketing managers
	Purchasing agents
	Technical service personnel
Research or engineering	Modification of existing products
	Application of basic research findings
	Experimental work with suppliers' new products
	By-product of work on other ideas
	Original thinking of recognized problem
Other company personnel	Suggestion systems
	Analysis of production processes and costs
	Analysis of maintenance and operating costs for existing products
	Analysis of markets for existing products and possible new products
Outside sources	Inventors and industrial designers
	Stockholders
	Management engineers, market research consultants, advertising agencies, and so on
	Purchase of an operating company
	Import or manufacture of products being exploited successfully in foreign countries
	Trade associations and government agencies

departments. *External* sources include inventers, patents, competitors, and consultants.

Most new-product ideas, however, occur within the company, according to a survey conducted by the Conference Board.* Internal sources of new product ideas arise from the research and development group, product managers, marketing personnel, and marketing research department.

Traditionally, new product ideas have come from research and development personnel, and in companies where the emphasis is on complex product technology, R&D personnel are relied upon for many of the new product ideas. Another important group generating new product ideas is, of course, marketing personnel since salespeople and technical sales personnel are constantly contacting customers and prospects. By determining what problems their customers have, the inspiration for many new products to solve these problems is generated.

One of the most fruitful ideas for generating new product ideas is the marketing research department. One of the functions of a marketing research department is to learn the needs of end-users or consumers of current products. Information of this type is obtained by the marketing research personnel from technical or research directors in consuming companies. Here the marketing research practitioner attempts to determine what the consumer wants in the way of new products and to obtain specifications for an "ideal" product. In addition, this person attempts to foresee the needs of the consumer that are not even anticipated at present. He or she also learns what competitive materials now satisfy these needs, at least in part, and then attempts to discover the specifications, advantages, and disadvantages of these materials. Using this information, he or she is then able to feed back information on consumer needs to the research director of the company to enable the director to more fruitfully orient the work of the laboratory research personnel.

In some companies combined marketing and R&D task forces have been set up to generate new product ideas. One of the advantages of the concept is that the personnel of both marketing and R&D are forced to consult regularly with each other. According to the Conference Board, at Engelhard Minerals and Chemicals Corporation, sales and technical personnel attack new product problems together in joint data-generating sessions. Out of one of these sessions grew the concept of Ultra-White 90, a kaolin-based, high-brightness coating clay, which has become one of the company's most successful new products.

Other company personnel used in generating new product ideas are product service specialists and other personnel having knowledge of the company's product line. The attempts of a number of companies to convert waste products into commercially saleable items are still another source of new product ideas. Examples cited by the Conference Board are:

*The material in this section is drawn from that survey, "Generating New Product Ideas," by E. Patrick McGuire, Conference Board Report No. 546, 1972.

1. The recovery of fly ash waste for subsequent marketing as a construction raw material
2. The recovery of various minerals and metals, such as nickel, cadmium, and others, from industrial effluents
3. The recovery of phenol chemicals from aqueous wastes of coke plants
4. The use of open-hearth steel slag to produce new building material aggregates

External Sources of New Product Ideas

New product ideas that originate outside the corporate environment generally come from two sources:

1. Primary sources, such as customers, distributors, consultants, suppliers, and competitors
2. Secondary sources, such as technical studies, newspaper and magazine leads, special market studies, and governmental reports

Still other new products result from the efforts of independent inventors or are obtained by the licensing of patents or technology devloped by other firms. These two subjects are discussed in Chapter 6.

Customers

Many companies solicit the suggestions of their customers and consider them to be a prime source of new product ideas.

Resellers

Industrial distributors and wholesalers are often called on to help with suggestions for new products. A manufacturer's distributors have a close knowledge of its products' users and uses, as well as a vested interest in the firm's new offerings.

Some firms make a special point of asking their resellers when they get a chance, "What are your customers asking for?" Beyond this, however, there is often no ready mechanism for eliciting new product suggestions from the resellers. Some companies have attempted to solve this by setting up "new-product advisory boards" made up of representatives from their distribution outlets, while others hold contests for their distributors to encourage submission of product recommendations.

A distributor for a chemical company suggested a method of inserting a low-cost polyethylene bag within the company's 55-gallon steel drums as a means of preventing corrosion between the chemicals and drum wall. The manufacturer was able to reduce its packaging costs significantly by shifting to the use of less expensive uncoated drums.

Most distributors are also in a position to help in the collection of competitive intelligence that might lead to new products.

Materials Suppliers

Suppliers of raw materials, equipment, and component parts often have technical expertise that permits them to suggest changes that result in new products for companies to which they sell. Also, the adoption of changes suggested by packaging forms can substantially alter the utility and image of a manufacturer's product.

New-Product Consultants

As is often the case with employing outside assistance, there are differences of opinion about the value of using new product consultants to recommend and develop product ideas. Some firms feel that such consultants are well worth their fees, while others say that the help given them by consultants has been minimal. The value of consultants will be discussed in more detail in Chapter 11, which is devoted to external sources of commercial development aid.

However, some firms have almost no technical staff; if they want assistance, they have to hire consultants. On the other hand, a number of firms with strong internal marketing and product development staffs routinely rely on consultants as well. Also, some executives favor the use of consultants because they can be hired for specific projects, only as needed, thus reducing the overhead and expense of product development.

Those managements who favor the retention of new-product consultants claim that the experience and detachment of these consultants is very important. Because of a consultant's expertise, he or she may be able to shorten product development schedules substantially.

Naturally, an advertising agency is not so much interested in selling a product idea as it is in obtaining new advertising business. But an agency knows that if it can develop an idea for a sound new product, the chances are excellent that it will get an account for it.

There are many situations, of course, in which a client would not expect to get new-product ideas from its advertising agency. This is usually the case where industrial products are concerned.

Marketing Research Agencies

In the course of research for a client, a marketing research firm will occasionally find promising segments of a market not being served, holes in the client's product line, or new uses for its products, just as would an internal marketing research department.

Venture Capital Firms

During the past decade, a number of financing firms have been organized that specialize in providing risk capital to newly formed manufacturing and marketing companies. Although these venture capital firms are concerned primarily with investing in new business ventures, their position also makes them a potential source of new product ideas.

Inventors contact such organizations seeking financial backing for their ideas. When such a course may not be the most practical method of capitalizing on an idea, the venture firm may suggest instead that the idea be licensed to an existing company.

For example, one venture firm was approached by a biochemist who had discovered a therapeutic substance derived from animal glands. The biochemist tried to persuade the venture firm to help him set up a small company to manufacture this biochemical. However, after a close look at the sales and profit potential of such a business, the venture firm recommended an alternative. It found a small producer of biochemicals who agreed to purchase a license for the newly discovered item.

Patent Attorneys

At times, independent patent attorneys function in the dual role of legal counsel and sales agent for their inventor clients. Often, a patent attorney's experience gives him or her a better understanding of where a new invention can best be marketed—that is, which companies are most likely to be interested in buying the invention that the inventor has. Some members of the patent bar reportedly act as "talent scouts" for manufacturers, keeping an eye open for promising ideas.

Retired Product Specialists

A number of companies sometimes hire retiring product specialists. One firm keeps track of the retirement not only of its own technical personnel but also of those in competitive firms. Once such persons have retired, the company's research director feels free to contact them about countinued service as new product consultants, with the reservation that the company will not solicit any confidential information pertaining to work done for their previous employers.

Product Designers

Outside designers, though usually retained by manufacturers for the purposes of refining product or package designs, are sometimes employed to help in generating new-product ideas.

In this latter role, designers may concern themselves not only with the configuration of a proposed product but also with the related matters of product use and preference.

Ideas from Competitors

In many cases, the introduction of a new product by a competitor causes a company to reexamine, or makes it become aware of, new product ideas that it may have overlooked.

Also, some companies are able to improve on their competitors' original ideas. Although the original company naturally has a competitive advantage by being first, it is still better to be late in the marketplace than not to be in at all. Also, if a company can move quickly enough, with an even better product, it is able in many cases to overcome the lead of the first company in the market.

Monitoring Competitive New Product Activities

Companies usually rely on field reports from the sales staff, trade journals, trade show exhibits, competitor promotional materials, and other sources to keep informed on the new products being introduced by their competitors. In addition, those selling through resellers also have them keep abreast of what is new in their markets.

Technology borrowed from foreign competitors sometimes proves valuable in developing new products for domestic markets.

Another valuable source of new product ideas is the individual inventor. As a matter of fact, although many people seem to believe that the day of the individual inventor has given way to the large industrial laboratories, there is evidence to the contrary. A number of studies have confirmed the continuing importance of the inventor. It has been estimated, for example, that during the period 1946-1955 two-thirds of the major innovations were the result of independent inventors and small companies.

The National Science Foundation estimates that most of the R&D dollars spent by industry are for product development rather than for basic research. The R&D dollars spent by industry are broken down as follows: product development, 78 percent; applied research, 18 percent; and basic research, 4 percent.

Inventors fall into two general categories: the professional full-time inventor and the casual or part-time inventor. The part-time inventor is reached through patent expositions, patent brokers, and patent attorneys.

Patent expositions are sponsored by professional companies that exist for this purpose and that provide display facilities or by patent brokers. Also, various state and federal agencies sponsor expositions. The U.S. Department of Commerce assists state agencies in this work through its Office of Invention and Innovation.

Patent brokers, who specialize in marketing inventors' ideas, operate in several ways. They may charge the inventor a flat fee or take a percentage of the royalties or at times purchase the invention itself.

Another method of generating new product ideas is the licensing route. The unused patents generated by companies in their research efforts can be useful to other companies. A number of companies have prepared lists of their surplus patents and have been successful in either licensing them to other companies or exchanging them for other patents that might be useful to them. The licensee gains the obvious benefit of a shortened product development time. For example, about 10 ten years ago, National Cash Register developed a microencapsulation process for use in carbonless paper. The process was licensed to other companies for use in perfumed paper, pressure-sensitive adhesives, and other applications.

Other new product ideas can be found by perusing trade journals, new product newsletters, and various technical books. Another important source is the U.S. government, which sponsors a great deal of technical research that is often available for use by industry. Also, a number of universities and nonprofit foundations support research that could be useful in generating new product ideas.

Bibliography

Articles and Books

Ansoff, H. Igor and Stewart, John M. "Strategies for a Technology-Based Business." *Harvard Business Review,* November-December 1967, p. 71.

Barasch, Kenneth L. *Marketing Problem Solver.* Fullerton, Calif: Cochrane Chase and Company, 1973.

Brion, John M. *Corporate Marketing Planning.* New York: Wiley & Sons, 1967.

Buell, Victor P., ed. *Handbook of Modern Marketing.* New York: McGraw-Hill Book Company, 1970.

Burick, Frank M. *How to Develop New Products for Sale to Industry.* Englewood Cliffs, N.J.: Prentice-Hall, 1971.

Corley, H. M., ed. *Successful Commercial Chemical Development.* New York: Wiley & Sons, 1954.

Giragosian, N. H., ed. *Chemical Marketing Research.* New York: Reinhold, 1967.

Hilton, Peter. *Handbook of New Product Development.* Englewood Cliffs, N.J.: Prentice-Hall, 1961.

McGuire, E. Patrick. *Generating New Product Ideas,* Report No. 546, The Conference Board, 1972.

Spitz, Edward A., ed. *Product Planning.* Auerbach, 1972.

Talley, Walter J., Jr. *The Profitable Product: Its Planning, Launching, and Management.* Englewood Cliffs, N.J.: Prentice-Hall, 1965.

Reports (All issued by the Conference Board)

Development of new products. A symposium on organization, with organization charts. *Organization for New Product Development.* EMM 11 (1966): 83 pp.

Independent inventor as a source of new product ideas. "The Casual Inventor." CBR, October 1971, pp. 54-57.

Industrial new products. "New Product Risk in Industrial Markets." CBR, November 1967, pp. 26-31.

Industrial products. Pretesting new-product ideas, estimating market potentials, and pretesting products and markets. Case studies are included. *Appraising the Market for New Industrial Products.* SBP 123 (1967):112 pp.

Innovations in new products. Pressures and challenges in marketing. "What Marketing Executives Want–New Products." CBR, May 1967, pp. 17-22.

Market segmentation. Various kinds of market segmentation and product implications are discussed. "The Implications of Marget Segmentation." CBR, March 1969, pp. 43-47.

Market testing. A symposium on concept testing, product testing, prototype testing, purchase simulation, and miniature market tests. *Market Testing Consumer Products.* EMM 12 (1967): 196 pp.

New product failures and some means to minimize them are discussed. "New Product Pressures." CBR, June 1971, pp. 16-24.

New product ideas. Internal and external stimulation of creativity and utilization of inventors and licensors. Examples of corporate waiver forms and policy statements. *Generating New Product Ideas.* Report 546 (1972):70 pp.

New products as a source and obstacle to corporate growth. A survey of opinion from the panel of senior marketing executives. *Sources of Corporate Growth.* EMM 24 (1970):7-9, 19.

"Patent Counsel in Industry." *Studies in Business Policy,* 112, 1964.

Planning responsibilities of product managers. *The Product Manager System.* EMM 8 (1965):126 pp.

Pricing. Surveys of business opinion and experience covering factors in pricing decisions, responsibility for setting prices and pretesting. "Pricing New Products." CBR, January 1964, pp. 7-14.

4
BUSINESS PLANNING

Henry Avery
USS Chemicals
Division of United States Steel Corporation
Pittsburgh, Pennsylvania

J. William McNeil
McNeil Associates
Pittsburgh, Pennsylvania

This chapter is intended as an overview of corporate planning. Top management's role in planning will be discussed, as well as the importance of comprehensive planning. The nature of business planning and plans are also treated.

There are many different facets of the planning function. This chapter will consider some of the most important. A brief description of the following types of planning will be included.

1. Marketing planning
2. Product planning
3. Financial planning
4. Diversification planning
5. Research and development planning
6. Planning in other functional areas

The process of developing plans will be described in the context of the network of corporate aims. A brief discussion of the nature and development of business strategies will be included but will be discussed more thoroughly in a later chapter. Also, this chapter will include a discussion of how to translate strategic planning into action.

Business Planning Shapes the Company's Future

What should be the nature of the company in the years ahead? What are the key assumptions that will be pertinent to the business over the next 10 years? What will be the leading or important external environmental forces at that time? How might they impinge or affect business deicisons being made today in regard to the company's present or anticipated business?

The business planning process forces the company to anticipate, develop strategy, and be concerned about its future. However, more importantly than being concerned about its future, it is forced to do something about it now!

It is the implementation of the business planning process that shapes the company's future.

Business Planning Defined

Business planning is concerned not only with the expansion or divestment of present businesses but with the identification and development of new businesses or ventures. Ideally, it identifies the company's business direction and provides the framework for making certain decisions that are vital and critical to the company's long-term health and survival.

Business planning càn be defined as the process of identifying, developing, evaluating, and selecting various alternatives; determining which alternative provides the most profitable opportunities; deciding what resources are needed; setting policies that will govern the development and use of these resources; and implementing action programs that will achieve the most profitable alternatives. It involves all functions in the company and integrates the planning of the individual functions so as to provide the company with its most profitable opportunities.

Early Origins and History

The origins of planning are obscure and ancient. Early man was forced to plan in order to survive. Planning was essential to the military and early explorers who sailed to distant lands many centuries ago. The voyage was their planning framework. They had to develop plans in order to survive and reach the eventual destination. Planning, although not new, has had a remarkable evolution over the centuries.

Corporate long-range planning, as we know it today, is principally a phenomenon of the twentieth century. Prior to 1900, business planning was both informal and intuitive. These two characteristics were most appropriate for the business organizations that existed at that time. Since that time, business has undergone major changes in size, complexity, and other characteristics that require that today's planning be more formal and comprehensive.

Prior to 1930, planning, as a subject, received little or no attention in the business literature of the day. This was evidence that the informal and intuitive approach to planning was much in favor. In a general way, the business economy was still somewhat simple and was oriented toward furnishing the basic needs of society. Considerable time was being devoted to operating management and the development of more efficient factory operations as the industrial revolution progressed. Although a few of the larger companies had some form of long-range planning, it was Henri Fayol who correctly evaluated this period as follows: "It has not yet become established custom to regard the staff as an organ of thinking, studying, and observation, whose chief function consists under managerial impetus, in preparing for the future and seeking out all possible improvements."

When the Depression of the 1930s developed, it forced many companies to devote all of their energies to taking what steps were needed to survive. Interest in formal long-range planning was expressed during this period, but not much progress was made in implementing it.

World War II is credited with being perhaps the largest single factor contributing to the development of formalized long-range planning. The war injected such concerns as quickly building military capacity, coping with a scarcity of resources, forcing abrupt technological change, and the need for involvement by the government in many activities of business. Perhaps the most overriding consideration was the substantial amount of planning carried on by the military in order to conduct the war. It was during this period that a certain amount of interchange in the process of planning took place between the military, the government, and business.

It was a natural progression that after the war, long-range planning would come into its own. Since 1950, this activity has received almost constant attention at management seminars. Most firms now have established planning activities. Again, a certain amount of this planning activity is informal and intuitive especially in the smaller firms. However, an overwhelming number of medium- and large-sized firms have initiated, designed, and installed formalized planning processes.

The past two decades have seen long-range planning recognized as an essential and integral part of the top management job. The growing complexity of business, the stepped up rate of change in technological advances, the number of new management skills that are constantly required, the rapid trend of business to become more global in nature, and the pressures of a changing external environment will require an increase in the quality and effectiveness of long-range planning in the years ahead.

History

1900	1970
Informal approach	Formal approach
Intuitive judgements	Technique and systematic analysis
Personalized method	Planning group, teamwork

Is It Needed?

The need for planning in all companies is not a temporary need but a continuous need. Unfortunately, many companies still look upon planning as something to be utilized primarily when the business is expanding and growing and something to be discarded when general business conditions slump or slide off.

When the business cycle turns down, many companies reduce their planning efforts or neglect planning entirely as the company scrambles for near-term profits. In reality, this is not the time to discard planning but rather to put more emphasis on near-term considerations. If we accept the premise that business planning is an integral part of the chief executive's job, it seems difficult for him or her to justify neglecting this phase of his or her responsibility, particularly when the company is on the down trend.

Solid, short-term planning can many times provide the necessary strategy for turning the company around and pointing the way to better operations. Perhaps the overriding reason why many companies neglect planning when the business cycle is down is because of its relationship with facilities and investment planning. Thus, a company may decide that if money is not currently needed or available for facilities, a planning effort is not needed. However, a comprehensive, overall business planning effort does not have the narrow function of facility planning. In fact, the installation of facilities is only a means for implementing the business plan. Aside from facilities, business planning is concerned with new markets, growth opportunities, new products, expansion of present markets, acquisitions, licensing opportunities, the creation of new patent positions, anticipating new competition, new processes, divestments, and so on. Thus, plant facilities are not ends in themselves but merely the means for carrying out all of the important elements and alternative decisions determined by the business plan.

The rate of change in today's and tomorrow's economy is expected to accelerate. Thus, the risk associated with developing and operating a business will be amplified. These key considerations establish planning as an important staff activity, as a basic function in the line operation, and as an integral part of the top executive's job.

Purpose of Planning

Business planning encompasses every facet of a company's business. It is involved with determining where the company should be headed and how to get there. Business planning identifies, analyzes, evaluates, and selects the alternatives that are open to the company.

The purpose of business planning is to determine the specific alternative or alternatives that will enable the company to capitalize on profitable opportunities with optimum results. Unfortunately there is no magic formula for achieving these results.

Business Planning is a Process

Business planning is a process that is *continuous.* It involves all of the functions of a company. It copes not only with the variables within the company but also with the ever-changing external environment.

The social, economic, political, and ecological forces in society can affect any company's plans. It is necessary that alternatives and assumptions be developed to provide for the changing nature of these external factors and adjust the company plan as needed. In order to meet these varied and changing situations, each company must customize or tailor the business planning process that best "fits" its own structure, organization, history, capital, products, markets, resources, and so forth.

Planning can be *informal,* particularly when utilized by companies that operate in a fairly stable technological environment, are not widely diversified as to products and markets, and are relatively small in their operations. Companies that are the converse of these factors, in varying degrees, usually require a *formalized* planning process. It is important that the process be customized to furnish the proper degree of formalization that is required. Formalization should not excessively proceduralize the planning but rather should make it an effective working tool whereby management can shape the company's future and meet the competitiveness of the marketplace and a changing environment.

As the rate of change makes itself felt, the complexity of operating a business can grow geometrically while the actual size of the company may be growing arithmetically. It is a company's planning process that narrows this "complexity gap."

Business Planning and "Performance Areas"

Business planning encompasses the key "performance areas" in business. Certain of these areas may also be defined as business functions. They are: marketing, production, technical, commercial development, products, markets and processes, raw materials, distribution, financial, future customer requirements, business resources and capabilities (that is, strengths and weaknesses), and future competition.

Many business people faced a much less complex environment in the earlier years of this century. Most businesses were relatively small and simple in operation; usually only a few of the "performance areas" just mentioned were pertinent. In addition, many of the entrepreneurs at that time were biased toward a single area such as marketing, production, or some other area. Today, if only a few areas of the overall business are considered in the planning process, the company can be seriously handicapped, and the company's business plan can become unrealistic and invalid.

Business Planning Within "Performance Areas"

All performance areas require and engage in some form of planning. The overall business planning function itself coordinates and integrates the planning inputs from the different performance areas. However, more importantly, it interfaces and interacts with them without being biased toward any single one of them—it optimizes the aggregate. Business planning is the unique tool that enables management to cope with all aspects of the business as opposed to single functions or performance areas.

Scope of Planning

It has been popular to think of the scope of business planning in terms of time—that is, short-term planning or long-term planning. Although time is an element in the planning process, scope is much more encompassing. For example, scope is involved when contrasting the future of a company's present businesses with the company's new, future businesses. It is the planning process that must continually determine the degree of emphasis that a company should place on present businesses versus new businesses. This emphasis changes with regard to the resources required, the results that can be anticipated, the health of the present businesses, the opportunities for new businesses, and so on.

A company may want to limit the business planning scope to present businesses because of limited resources. In this kind of situation, planning may be more properly termed production capacity planning, interspersed with marketing planning. Thus, the planning is more a function of expansion and extension of present business activities. This situation may be considered as a "holding action" as most businesses are constantly changing as products die or are replaced and markets change in regard to the needs of the customer.

Because new businesses often require new human, financial, and physical resources, there can be a conflict between a desire to strengthen present positions as opposed to developing new positions. The strengthening of a present position may "choke off" resources for new business areas or merely extend the company's participation in an area that is going downhill.

Determining the planning scope that best fits its needs is an important part of a company's planning effort.

Coping With the Environment

Most of the business functions or performance areas mentioned previously are concerned with an overall environment that includes both internal and external factors.

Internal

One factor that is not controllable is internal attitudes. Unfortunately many attitudes within companies are rigid. Unless altered or changed, such attitudes can exert a strong negative influence on the development of business planning.

External

The external environment is constantly changing, and the rate of change is expected to accelerate for many years. One of the newer tools used for gaining better insights into the constantly changing external environment is technological forecasting. This tool is gaining wide acceptance as the need for its application grows within industry. Obviously every business makes a technical forecast whether they call it by that name or not. Some of the more formal specific methods used in technological forecasting include the well-known Delphi method, extrapolation, trend analysis, and others. Technological forecasting usually refers to predictions in the technical area of the business or an industry. However, when we consider all of the critical and important planning elements that make up a company's operations, as previously mentioned, we need to be concerned with what might be called "change forecasting," which, although similar in nature, applies not only to the technical planning area but to all of the other key planning areas of the business. The utilization of this predictive approach across the business is compatible with Peter Drucker's definition of planning: "Planning is concerned with the futurity of present decisions."

Business Planning

- Internal
  - Attitudes
- External
  - Technological Forecasting
    - Predictions in the technical area of the business or industry
  - "Change Forecasting"
    - Applies to all of the performance areas of the business

Planning and the Computer

The computer is a prime tool in business today and will be so in the future. However, the actual rate of acceptance is dependent on how fast the knowledge of management proceeds in understanding the capabilities and proper use of the computer. Business information systems continue to become more essential as part of a company's competitive capability and can be a key element in supporting management in decision making. Business planning decisions, as part of the chief executive's responsibilities, can be strongly influenced by such systems.

Management science, linked with business information systems, can be expected to play an expanding role in business planning in the years ahead. As we experience a wider utilization of the computer coupled with a broadened acceptance by industry of the application of mathematical models, there will be new uses made of scientific methods of analysis. The use of advanced scientific method for business planning will accelerate as industry management gains a better understanding of this approach and the contribution it can make to better decision making. It is unfortunate that industry management has often considered scientific method as an approach that is an alternative to human judgment. Instead, they are interrelated and complementary. When properly used together, better decisions can be anticipated.

Planning and Profitability

Profits don't happen; they're planned. Profit opportunities must be identified, analyzed, and evaluated. They must be aggressively pursued in order to be accomplished. Threats to the business must be recognized so that they can be avoided or coped with as they arise. Decisions pertaining to equipment, technnology, markets, raw materials, people, finance, and so on, affect profits. If the right decisions can be made at the right time, there is a good chance of achieving maximum profit. The business planning process seeks to do this continuously. It determines and evaluates a company's resources and capabilities in order to optimize the financial returns that can be accomplished through the effective utilization of the resources and capabilities.

Business Planning

Focuses on Profit Growth

- Should all the present businesses be expanded?
- Are mature businesses in a "holding pattern"? Curtail?
- Are certain businesses really worth additional investment?

The Business Planning Group—Who, What, Why?

A company's *business planning group* consists of the key people who make up top management. This group, under the direction of the chief executive, assists him or her in discharging his or her basic responsibility for the development and execution of the company's business plan. The group's effective collaboration and consensus can provide the support that the chief executive needs in order to ensure the success of the eventual business plan that is conceived.

The exact manner in which the business planning group functions will vary from company to company. Again, a customized approach is required that best fits each particular organization. The chief executive will generally require the assistance of a moderator who will moderate the business planning group through the planning process. The moderator may or may not be selected from among the membership of the business planning group. However, the overriding consideration in the selection is that the moderator be an individual with a strong knowledge of planning methodology and technique. The other members of the planning group will not be as concerned with methodology and technique but rather with the input of sound business judgments into the company's business plan.

Function of the Business Planning Group

The business planning group must arrive at a strategy by identifying the overall alternatives and evaluating them—by determining their risk levels and profitability, the key actions needed, the resource allocations required, and the timing—all within the perspective of ties with the past, a grasp for present problems, and judgments about the future.

Business Planning Group

Must Identify

- Overall Alternatives
- Risk Levels
- Key Actions Needed
- Resource Allocations Required
- Timing Profitability

Within Perspective of:

- Past
- Present
- Future

Master Strategy Planning

The term *master strategy plan* refers to the overall company plan that has integrated all of the functional or "performance area" plans such as the marketing plan, the research plan, and so on.

There are very few established procedures and techniques that can be universally applied to strategy planning. As a result, planning cannot be reduced to simple procedures or systems. The process itself is complex, dynamic, and constantly changing, just as companies change under the influence of changing internal and external environments. It is this situation that tests the skill of anyone engaged in planning. Thus, methodology must be developed and utilized for each situation so that a company's planning effort, and the results obtained, can be optimized.

Over the years a considerable amount of literature has accumulated on strategy planning. However, a vacuum exists in regard to the techniques or tools needed to select the most desirable alternative or strategies required for a company's profit growth. The breakthrough in master stragety planning is not entrepreneurial concepts but in work methods, analysis, and particularly technique. This breakthrough is not widely used as yet, but it is gaining acceptance. Management is still in the early conceptual phase of strategic planning. However, techniques, classification systems, methods, and so forth, are not standing still. They are constantly evolving and can be expected to continue to undergo rapid change.

The technique concerned with developing the company's master strategy plan is quite different from budget technique or forecasting technique, product planning technique, facility planning technique, or goal setting. Unfortunately, it is no longer adequate to set goals and merely work hard to achieve them. Goal setting alone is a quantitative approach. There is a real need for technique that leads away from placing a heavy emphasis on numbers and forecasts. What is needed is a more qualitative, judgmental approach. Although numbers and forecasts are necessary, it is technique that can put the numbers and forecasts into a different perspective, causing people to think in different ways.

Planning Techniques

Need for Techniques that

- Develop the direction of the company
- Lead away from heavy emphasis on numbers and forecasts
- Can put numbers and forecasts in perspective

Identifying Alternatives

The development of the company's master strategy plan requires that the business planning group identify overall alternatives that are feasible and achievable. Such alternatives could include the acquisition of a new business, the invention and development of new products in research, the expansion of a present business or replacement of old facilities, divestment, licensing certain products, moving into different market areas or creating new markets, and so forth. Two or more of the single alternatives may be interrelated and lend themselves to being combined into a single alternative. One or more of the single or combined alternatives that are identified will need to be judged as having the best chance of contributing to the company's profitability and competitiveness in the future, hopefully with the least risk and the most profit growth. Many alternatives are open to certain companies, whereas others will have rather few possibilities, usually due to a lack of resources.

Technique is used to identify the various alternatives that must be eventually developed and evaluated. It forces decisions as the business planning group proceeds and heads off indecisiveness. Thus, concensus among the members of the business planning group is constantly being achieved, through the guidance of the moderator, as the planning moves ahead. Technique also stimulates "guided" idea generation leading to the identification of possible alternatives.

Identification of Alternatives

Need to identify various alternatives that are feasible to accomplish

Technique

Used to identify various alternatives

Defining and Structuring the Alternatives

After identifying the alternatives, the business planning group must define and structure them. The structuring of each alternative is expressed in terms of products, processes, and markets. Such structuring defines the preliminary scope for each alternative. As the structuring technique proceeds, it is possible to "screen out" certain alternatives that show little chance of success. Such "screen outs" result from the judgment of members of the business planning group rather than from detailed analyses. The alternatives that survive are then fully structured, as needed, in preparation for evaluation.

Defining and Structuring Alternatives

Alternatives are structured in terms of scope expressed as products, processes and markets.

Certain alternatives can usually be "screened out."

Alternatives that survive must be readied for evaluation.

Evaluating the Alternatives

The evaluation step also involves technique. However, the technique does not rely on the development of fine-line quantitative studies and a voluminous collection of facts. Rather it is almost completely dependent on the multiple and diverse judgments given as inputs to the evaluation by the members of the business planning group. Certain questions and criteria applicable to the nature of the particular alternatives being evaluated must be developed and used as the framework for the evaluation. These criteria encompass the "performance areas" mentioned earlier. It is really the recognition of the "performance areas"—and the use of them—that forms the foundation for the company's master strategy plan.

Evaluation of Alternatives

Not focused on fine-line quantitative studies.

Rather

Focused on multiple and diverse judgments drawn from business planning group.

Selecting the Alternatives—A Critical Step

Once again technique comes into play in selecting the final alternative or alternatives that will form the basis of the company's master strategy plan. As with the evaluation step, the essential input is the collective judgment of the members of the business planning group. Optimally, these judgments lead to optimum profit growth opportunities within a framework of reasonable risks. These decisions include entry into new business areas as well as expansion of present businesses, divestments, licensing opportunities, acquisitions, penetration of new markets, and so on. The risk levels applicable to each alternative must be gauged with a solid understanding of the company's present business and competition and a good knowledge of the present and future internal and external environment that must be met. It is this technique of selecing the "best" alternative that provides the management tool that makes the achievement of optimum results possible.

Selection of Alternatives

Selection of alternatives by business planning group affect

- Entry into new business areas
- Expansion of present businesses
- Divestments
- Risk levels to be taken

Commercial Development and Business Planning

As noted earlier, commercial development is one of the key performance areas in business planning. It is an integral part of the business planning process along with other key functions in the overall business. Commercial development must be involved in all aspects of the company's business planning so that its implementation role in carrying out the planning can be most effective. This involvement would apply not only to the expansion of present business areas but particularly to the identification and establishment of new directions and new businesses by the company.

Business planning, although a major responsibility of the top executive, involves and cuts across all parts of the company's business. Thus, to truly accomplish overall business plannig, it is necessary that inputs be made by all functions within the company. The inputs made by the commercial development function can be most significant.

The commercial development activity, if it is to be effective, must be knowledgeable about, and involved with, the overall scope of the company's business. This scope is a broad one when related to other functions of a more specialized nature—for example, sales and marketing, research, and so on. It is this breadth of scope required of commercial development that qualifies it as a general management function rather than a specialty function. This is not to imply that everyone engaged in commercial development is a general manager or a potential general manager. However, it does imply that the function must provide the general management capability either directly or by drawing on other people or functions within the company.

Inputs made by commercial development into the business planning effort need to be heavily futures oriented, particularly if the company's objectives focus on growth. It is a key focal point for identifying and evolving new business growth opportunities and in shaping and implementing the company's objectives —particularly growth. Its practitioners must not only be sensitive to change but must promote change!

Commercial development in the chemical industry has changed markedly in the past two decades. Increased risks, the emergence of global competition, the information explosion, the stepped up complexity of industry, the accelerated rate of change of social, economic, political, ecological, and technological factors are some of the trends that are requiring companies to strengthen their development efforts. Thus, companies are staffing their programs with people who have many diverse skills and capabilities. Some of the major elements of the commercial development activity that support the company's business planning effort include: facilities planning; overall product planning (new and existing products); economic and technical forecasting; overall marketing strategies (present and new markets); product, process, and raw material development; analyses of the company's strengths, weaknesses, resources, and capabilities; environmental and marketing analyses; and venture analyses. As various growth alternatives are developed in the planning process, commercial development can be an effective arm for evaluating and analyzing the identified alternatives. In addition, commercial development can recommend the criteria for the selection of alternatives and recommend alternatives that seem most likely to achieve the company's overall objectives.

Installing the Business Planning or Commercial Development Functions

The importance of designing a planning process for a company cannot be overemphasized. Each company requires a somewhat different design to meet its needs and goals. Although the planning design is critical, nothing happens until the planning process is finally established and installed in the the company. It is critically dependent on the attitudes and cooperation of certain key management people in the company who must work within the overall planning process. Strong support of planning by the chief executive can do much toward bending these attitudes but often will not completely change them. This opposition, which often arises, usually originates with persons who may feel threatened by the planning process or who do not fully understand planning. They may feel uncomfortable when dealing explicitly with matters in the future that cannot be dealt with on the basis of facts. Some may feel that their managerial prerogatives are being encroached upon. Many responsible people are reluctant to commit themselves to forecasts of future events. As a result, they are reluctant to participate in planning activities. A similar kind of opposition can also apply to the establishment and installation of the commercial development function in the company.

Because many attitudes and practices are usually so rigid and entrenched, it is very difficult to establish and install business or commercial development quickly. In fact, it may require several years before an effective planning process can be installed. Hopefully, over this period of time, enough enlightenment concerning business planning or commercial development can be brought to bear so that the people involved can understand the purposes and benefits. However, ideally, these people will not only understand but will make the necessary contributions to ensure that the business planning or commercial development activities are effective.

When business planning or commercial development is new to a company, a plan must be devised for making the installation. Ideally it should be a step-by-step installation. The speed of the installation should be governed by the acceptance rate of those participating and the nature of the results being achieved. Much patience and perseverance are required. If the installation moves too slowly or too quickly it can die. Thus, *timing, sequencing,* and *pacing,* so critical to any planning process, become highlighted as key factors in the installation. Once an installation is established, it requires a continuous renewal and updating because of turnover in the organization, different needs or objectives being developed by the company, and the new planning techniques and methods that are constantly being developed.

Bibliography

Planning (books)

Ackoff, Russell L. *A Concept of Corporate Planning.* New York: Wiley-Interscience, 1970, 157 pp.

Ansoff, H. Igor. *Corporate Strategy: An Analytic Approach to Business Policy for Growth and Expansion.* New York: McGraw-Hill, 1965, 241 pp.

Ansoff, H. Igor, Declerck, Roger P., and Hayes, Robert C. *From Strategic Planning to Strategic Management.* New York: American Management Association, 1976.

Anthony, Robert N. *Planning and Control Systems: A Framework for Analysis.* Boston: Harvard Business School, 1965, 180 pp.

Argenti, John. *Corporate Planning: A Practical Guide.* London: George Allen and Unwin, 1968, 272 pp.

Berg, Thomas L., and Shuchman, Abe. *Product Strategy and Management.* New York: Host, Rinehart & Winston, 1963, 610 pp.

Blecke, Curtis J. *Financial Analysis for Decision Making.* Englewood Cliffs, N.J.: Prentice-Hall, 1966, 200 pp.

Bower, Joseph L. *Managing the Resource Allocation Process: A Study of Corporate Planning and Investment.* Cambridge, Mass: Harvard University, 1970, 363 pp.

Branch, Melville C. *The Corporate Planning Process.* Cambridge, Mass: The Riverside Press, American Management Association, 1962, 253 pp.

——. *Planning: Aspects and Applications.* New York: Wiley, 1966, 333 pp.

Cannon, J. Thomas. *Business Strategy and Policy.* New York: Harcourt, Brace & World, 1968, 571 pp.

Collier, James R. *Effective Long Range Business Planning.* Englewood Cliffs, N.J.: Prentice-Hall, 1968, 187 pp.

Drucker, Peter F. *The Age of Discontinuity.* New York: Harper & Row, 1968, 401 pp.

——. *Managing for Results: Economic Tasks and Risk-taking Decisions.* New York: Harper & Row, 1964, 240 pp.

Ewing, David W. *The Human Side of Planning.* New York: Macmillan, 1969, 211 pp.

——, ed. *Long-Range Planning for Management.* New York: Harper & Row, 1964, 565 pp.

——. *The Practice of Planning.* New York: Harper & Row, 1968, 149 pp.

Ferrell, Robert W. *Customer Oriented Planning.* New York: American Management Association, 1964, 145 pp.

Goslin, Lewis N. *The Product-Planning System.* Homewood, Ill.: The Irwin Series in Operations Management, 1967.

Henry, Harold W. *Long-Range Planning Practices in 45 Industrial Companies.* Englewood Cliffs, N.J.: Prentice-Hall, 1967, 161 pp.

Hilton, Peter. *Handbook of New Product Development.* Englewood Cliffs, N.J.: Prentice-Hall, 1961, 223 pp.

Hughes, Charles L. *Goal Setting: Key to Individual and Organizatinal Effectiveness.* New York: American Management Association, 1965, 159 pp.

Hussey, D. E. *Introducing Corporate Planning.* Elmsford, N.Y.: Pergamon Press, 1971, 210 pp.

Irwin, Patrick H. *Business Planning: Key to Profit Growth.* Toronto: The Ryerson Press, 1969, 167 pp.

Katz, Robert L. *Cases and Concepts in Corporate Strategy.* Englewood Cliffs, N.J.: Prentice-Hall, 1970, 820 pp.

LeBreton, Preston P., and Henning, Dale A. *Planning Theory.* Englewood Cliffs, N.J.: Prentice-Hall, 1964, 357 pp.

Linowes, David F. *Managing Growth Through Acquisition.* New York: American Management Association, 1968, 192 pp.

Lorsch, Jay W. *Product Innovation and Organization.* New York: Macmillan, Columbia University, 1965, 184 pp.

Martin, Elizabeth, ed. *New Products/New Profits: Company Experiences in New Product Planning.* New York: American Management Association, 1964, 303 pp.

Miller, Ernest C. *Objectives & Standards: An Approach to Planning and Control.* New York: American Management Association, 1966, AMA Research Study no. 74, 120 pp.

Newell, William T., Jr. *Long-Range Planning Policies and Practices.* Austin, Tex: Bureau of Business Research, University of Texas, 1963, 199 pp.

Payne, Bruce. *Planning for Company Growth: The Executives Guide to Effective Long-Range Planning.* New York: McGraw-Hill, 1963, 316 pp.

Presanis, Aris. *Corporate Planning in Industry.* London: Business Publications Limited, 1968, 167 pp.

Rothschild, William E. *Putting it All Together.* New York, American Management Association, 1976, 262 pp.

St. Thomas, Charles E. *Practical Business Planning: A Manager's Guide.* American Management Association Handbook, The Haddon Craftsmen, Inc., 1965, 141 pp.

Scott, Brian W. *Long-Range Planning in American Industry.* New York: Colonial Press, 1955, American Management Association, 288 pp.

Simmons, John D. *Long-Range Profit Planning, Planning Your Profits for the Year Ahead.* NAA Research Report 42, National Association of Accountants, 1964, 111 pp.

Steiner, George A., ed. *Managerial Long-Range Planning.* New York: McGraw-Hill, 1963, 334 pp.

——. *Top Management Planning.* New York: Macmillan, 1969, 794 pp.

Steiner, George, A., and Cannon, Warren M., *Multinational Corporate Planning.* New York: Macmillan, 1966, 330 pp.

Talley, Walter J., Jr. *The Profitable Products: Its Planning, Launching, and Management.* Englewood Cliffs, N.J.: Prentice-Hall, 1965, 199 pp.

Wallenstein, Gerd D. *Concepts and Practice of Product Planning.* New York: American Management Association, 1968, 125 pp.

Warren, E. Kirby. *Long-Range Planning: The Executive Viewpoint.* Englewood Cliffs, N.J.: Prentice-Hall, 1966, 108 pp.

Watton, Harry B. *New-Product Planning: A Practical Guide for Diversification.* Englewood Cliffs, N.J.: Prentice-Hall, 1969, 176 pp.

Ward, E. Peter. *The Dynamics of Planning.* Canada: Pergamon Press, 1970, 347 pp.

Planning (articles)

"Formal Planning—The Staff Planner's Role at Start-Up." Stanford Research Institute Report no. 250, July 1965.

Gilmore, Frank F., and Brandenburg, Richard G. "Anatomy of Corporate Planning." *Harvard Business Review,* November-December 1962.

Henry, Harold W. "Formal Long-Range Planning and Corporation Performance." *Michigan Business Review,* November 1968.

Kami, Michael J. "Planning: Realities vs Theory." NICB Record, *Managing Change,* January 1967.

Marsy, William F. "Planning in Marketing" (a selected bibliography). Cambridge, Mass.: The M.I.T. Press, 19

Mason, R. Hal. "Developing a Planning Organization." *Business Horizons,* August 1969.

"Perspectives on Corporate Strategy." The Boston Consulting Group, Inc. Boston, Mass.

"Strategic Planning at Air Products," James F. Dempsey, Air Products, speech delivered before the Chemical Marketing and Economics Division of the American Chemical Society, Chicago, Illinois, August 30, 1977.

Thompson, Stewart. "How Companies Plan." AMA Research Study no. 54, American Management Association.

Technological Forecasting (books)

Bright, James R. *Technological Forecasting for Industry and Government.* Englewood Cliffs, N.J.: Prentice-Hall, 1968, 484 pp.

——, ed. *Technological Planning on the Corporate Level, Proceedings of a Conference Sponsored by the Associates of the Harvard Business School,* September 8-9, 1961, Cambridge, Mass.: Harvard University Press, 1962, 253 pp.

Cetron, Marvin, J. *Technological Forecasting: A Practical Approach.* New York: Technology Forecasting Institute, a subsidiary of Gordon and Breach, 1969, 343 pp.

McHale, John. *The Future of the Future.* New York: Braziller, 1970, 322 pp.

Technological Forecasting and Planning (journals)

Futures, The Journal of Forecasting and Planning, Published by IPC Science and Technology Press Ltd., UK, quarterly, in cooperation with The Institute for the Future, U.S.A., Editor: Guy Streatfield (Surrey, England), U.S. address: Futures, IPC (America) Inc., 300 East 42nd St., New York, N.Y. 10017.

The Futurist, a Journal of Forecasts, Trends and Ideas About the Future, published bimonthly by the World Future Society, P.O. Box 19285, Twentieth Street Station, Washington, D.C. 20036.

Long Range Planning, The Journal of the Society for Long Range Planning, Editor: Bernard Taylor (England), Published by Pergamon Press (England). U.S. address: Fairview Park, Elmsford, New York 10523.

Technological Forecasting, An International Journal, American Elsevier Publishing Company, Inc., 52 Vanderbilt Ave., New York, N.Y. 10017.

5

NEW VENTURE MANAGEMENT

William Copulsky
W. R. Grace & Company
New York, New York

Introduction

One of the continuing problems of growing companies has been the generation of additional sales through new products and new ventures. However, in recent years there has been a shift in focus as to the nature of the problem. This shift in focus is reflected in the description of the process in two books co-authored by the author of this chapter over a 26-year period. In 1948, in *The Marketing of Chemical Products*[1] a chapter was devoted to commercial development of new products, and the entire focus was on formalized procedures. In 1974, when the book *Entrepreneurship and the Corporation*[2] appeared, the entire focus of the chapter dealing with new ventures was on *people.*

A body of mythology is bound is arise around new ventures and their management. Articles in the trade press are generally public-relations-oriented. Successes may be exaggerated or publicized prematurely. Failures are less often publicized, and, even when they are, details are lacking. The cumulative costs of new-venture activity are not generally available. We hear about the successes and what they return but rarely about the accompanying failures and their costs and losses relative to the profits of the successes.

On the other hand, the magnificent failures such as Corfam or the Edsel are overstressed. We forget the many successes of the same companies. Ford has had many outstanding successes but is not quite forgiven for the Edsel.

Although Du Pont had a highly publicized $100-million failure in Corfam, a leather substitute, they also had by the end of 1973 a group of 24 new products targeted for profitability. These included Qiana, a silklike synthetic fiber, Nomex, a flame-resistant polyamide fiber, a Freon food-freezing system, and a line of electronic instruments. This group of new products netted about $71 million on $500 million of sales in 1973.[3]

Defining Terms

One way to start thinking about the management of new ventures is to define the terms. What do we mean by such words as management, new, and venture?

Management deals with the control of things and persons through the use of skill and care with a goal of successfully controlling the course of events. It is part science, part art—science in the sense that we have accumulated a body of organized knowledge about management, and art in that there are many elements of the process that seem incapable of transfer from one person to another. The word "control" is one element of the mystery. Who controls? Who is controlled? What is controlled?

A *new venture* is a business that has arrived at the point where management has decided that the product or process has a good chance to be successful commercially. This means that (1) it is marketable, (2) can perform technically as required (3) can be made or sold profitably at a return matching the corporate goal, and (4) can grow to a size matching the corporate goal.

The business idea that is the base for the new venture can come out of the company's own research or development or can be based on purchased know-how or could be acquired as a going new business that is still in an early stage.

Management As One Factor in New Venture Success

Management problems are "people problems." Good management is only one of the factors in the success of new ventures, but it is one which is receiving increasing attention as a factor of prime importance.

President Nixon in his message to Congress on science and technology in 1972 listed some of the factors involved:

> . . . scientific discovery alone is not enough. Even the most important breakthrough will have little impact on our lives unless it is put to use—and putting an idea to use is a far more complex process than has often been appreciated. To accomplish this transformation, we must combine the genius of invention with the skills of entrepreneurship, management, marketing and finance.[4]

A recent survey[5] indicates that business people and others concerned with commercializing innovations see management problems as second only to marketing problems in listing barriers to successful commercial innovation. Industrialists, financial managers, government officials, labor leaders, and college and university academicians all tended to this conclusions. Other important barriers were identified: finance, organizational problems, lack of seed capital, and the effect of depressive governmental policies and practices (patents, antitrust, and other regulations). It is interesting that technology itself was not a barrier to technological innovation. Our scientists are capable of inventing whatever the system can sell—but the system cannot sell anything that is invented, at least not at the particular time that the invention comes out of the laboratory.

But closely following market-related problems as barriers to new products are management "people problems" related to corporate organization and behavior such as:

1. The high risk of an individual being blamed for failure
2. The threat to the established corporate hierarchy
3. The problem of using highly functionalized specialists to act in an entrepreneurial way requiring knowledge and the ability to handle many business functions

The "people problem" starts with top management. Part of the problem would appear to be rationally inspired. Management must be convinced that existing products will not yield the return on investment or growth needed to meet corporate goals, and therefore new products or investments are needed to supplement internal growth. In addition, top management must have reasonable goals for new projects in terms of (1) the expected return, (2) the time period to reach profitability, and (3) the eventual size of the new venture.

However, there are also less rational considerations. Dr. Mortimer Feinberg has pointed out that innovation is always a source of conflict.[6] Although all are for innovation in the abstract, all are against innovations when they become reality. Real innovations disturb employees who have grown accustomed to their position and status in the corporate hierarchy. When disruption is threatened by innovation, employees react by resistance to the innovation. This means that within the corporate framework, a new successful innovation must not only be technically sound and fit marketing needs but must also be carefully and correctly introduced internally. If not, progress will be defeated by internal conflict.

Although management surveys as the one described list marketing problems first and "people problems" second as barriers to successful commercial development of new innovation, there is a growing perception by students of the subject that people are the number one element in the innovation process.

The management of innovation may be more important than innovation itself when the goal is profit. Successful entrepreneurs often have a path to profits that resembles a random walk. Many innovations have their commercial successes in applications not originally conceived by the inventor. Many radical innovations are in areas in which the innovating company is not active and has little strengths in management, marketing, manufacturing, or technology. Such radical innovations would not result from conventional long-range planning. When such technical innovations occur, the successful company will need management innovation to exploit the possibilities.

Management–Scientific and Entrepreneurial

In using the term "management," one must recognize many types and varieties of management. As companies grow in size, they progress from "entrepreneurial" to "scientific" management. Entrepreneurial management is more typical of the small company. It is management that can encompass the factors of production and get them to work together to make profit. As the company grows, management functions become more specialized and management becomes more "scientific." The scientific manager can devote time to understand his specialty in greater depth. He may use complex mathematical models. He may be superior in operations of a repetitive nature where historical patterns predominate. But the entrepreneur will prove superior in unstructured new situations where he has to proceed by feel, much as does the grand-master of chess who can beat a chess-playing computer that knows only history, but not really "how to play" chess.

The development of new products seems more successful in small companies and small organizations than in large corporations. The essence of the small company is the entrepreneur, the person who can manage the other factors of production–land, labor, and capital–to produce profit. He has freedom of action to get results. Observing the success of small companies in entering new businesses, many large corporations organized "new ventures" as autonomous units within their normal structures to overcome people problems.

In a large corporation, freedom of action has to be less than that of an entrepreneur starting his own business, because top management has a hand in all business decisions.

However, some of the attributes of the small-company entrepreneurial situation might be fitted in within the large-company framework. For example: (1) rewards related to results, (2) quick feedback of results, and (3) assumption of personal sacrifice or risk by the corporate entrepreneur.

At the same time, new ventures should be separated from the management of existing products. Keeping new-venture management separate from managing established products is almost a necessity to (1) assure enough time is spent on new products to insure success and (2) match an entrepreneurial atmosphere to entrepreneurial-type personnel.

These concepts are not completely new. As early as 1959, Colgate-Palmolive established "invention groups" that included marketing research, production, and laboratory research personnel to improve the chances of successfully developing new products.[7]

Small teams not only furnish entrepreneurship but allow team members to seek quickly the results of their individual efforts and to "put themselves in the shoes" of the other team members. Mort D. Margolis, president of the Autonetics Division of Rockwell International, Anaheim, California, puts his new project teams into one area. "We told our project manager that he was now in business for himself—just as if he had his own small company."[8] As a result, says Margolis, all personnel could see their effects on output. Each saw a personal challenge in meeting the needs of the team. They talked to each other instead of writing memos. They began to understand each other. Eye to eye, they learned respect for each other's expertise.

Douglas McGregor has pointed out that teamwork is superior to rigid vertical bureaucratic lines of authority in accomplishing goals.[9] This is especially true when the members of the team have to be technically proficient and must act as professionals.

Such small teams can also be highly motivated when (1) they can influence the outcome of events, and (2) they are not controlled by a rigid long-term plan. This has been pointed out in a recent study by Earl R. Driver: ". . . individuals who perceive themselves as able to influence the outcome of events are more likely to be associated with entrepreneurial activity, than individuals who feel that their personal actions have little influence on the outcome of events."[10] Detailed long-range "top-down" planning may discourage entrepreneurship. "Entrepreneurial activity is likely to be low when the mission of the organization is well defined for a number of years into the future."

But the people as well as the organizational structures have to be entrepreneurial. Success of a new venture in a large corporation will have to depend on a committed entrepreneur with freedom to fail. The key to new ventures may be in finding such a person. Too often the new venture manager is the wrong type—the trustworthy conventional manager, the manager over his or her head in a new venture, the manager who lacks the staying power, the deal maker, the mathematically oriented M.B.A. The right person is rare, especially in a large corporation.

In large corporations, there may be a number of new ventures coming to the forefront at one time. In such a case, there will be a number of new venture teams under some organizational umbrella. The corporation is faced not only with the problems of each new venture but with managing several new ventures. New questions arise: (1) How many new ventures of what size can be managed at the same time? (2) How should failures and liquidations be handled?

Entrepreneurship as the Key

Although entrepreneurship has been stressed as the key to success of new ventures, it is only poorly understood, especially by management of large corporations. Entrepreneurship is the spark of small-company operations, but introducing it into a large corporation has been described as being similar to "asking an elephant to ice-skate." A knowledge of entrepreneurship can be very advantageous to a large corporation in handling new products and ventures.

Entrepreneurs have generally been born, not made. Some have both entrepreneurial qualities and a rare combination of other qualities that make them more adaptable to corporate life than the classical entrepreneur would be. Although recent work indicates that perhaps entrepreneurs can be trained or that inherent entrepreneurial qualities can be upgraded, these cannot be accomplished by conventional business education. The M.B.A. course at a graduate school is not particularly entrepreneurially oriented. Entrepreneurs need a small company or simulation of the small-company atmosphere to thrive. Conditions that kill entrepreneurial spirit must be avoided. Unfortunately the very rigid organizational structure of the large corporation is antientrepreneurial. Modifications are needed. Such modifications include "new venture" organizations, or corporate spin-offs. The 3M Company is a prime example of how entrepreneurship can successfully and profitably develop new ventures in a large corporation. The 3M Company is essentially a collection of many small new ventures. It takes many good bricks to build an edifice of strength. Essentially such organizations make "every employee an entrepreneur." They combine the advantages of small-company entrepreneurship and corporate bureaucracy.

The understanding of the entrepreneurial factor, especially in the large-company atmosphere, is far from complete. Corporate management has much to learn about how to identify entrepreneurs, how to train them, how to keep them happy through use of proper motivations, and how to organize in such a way as to optimize entrepreneurial qualities. Nonentrepreneurial personnel are still often selected for entrepreneurial tasks.

Routes to New Venture Management

Introduction

As has been described, success in new ventures within the framework of the large corporation appears to depend on the ability to simulate the small-company atmosphere. Entrepreneurship stands for an ability to combine production factors to make profit and highly self-motivated people. Large corporations have developed various organizational devices to simulate and stimulate the right climate for profitable commercial innovation.

Various routes to new venture management include the following: (1) A new venture management department, (2) task forces assembled only as needed, (3) separate subsidiaries handling internal new ventures or outside venture capital investments, and (4) spin-offs.

New Venture Management Departments

One popular route has been the formation of a separate department to manage new ventures.

Of the various available routes to new venture management, the new venture department has been the least viable because it was relatively inflexible and was not isolated from overall corporate problems. In addition, problems of adequately motivating and compensating the corporate entrepreneurs were never satisfactorily solved.

In 1970, "venture management" was the fad of the year. It was promoted as *the way* commercially to develop new products and businesses in the large corporation and keep "corporate entrepreneurs" happy. Almost one-third of the largest U.S. companies likely had a formal venture-management function. By 1974, very few did. There were a number of reasons for the rapid decline of the venture management fad: (1) the recession, (2) overvisibility of the function, and (3) antagonism of pressure for quick returns.

The typical venture-management department spent a great deal of effort on formalized procedures ranging from checklists to sophisticated mathematical models. Often, too, much time was spent on setting up the departments, its organization, and procedures. The new ventures themselves never had a chance. Two years after its start the new department was finally organized—in time to be liquidated because of its inability to commercialize new ventures.

In other cases, new venture departments were a way of liquidating expensive research efforts. The "research" departments were converted to "new venture" departments. Becoming visible and measurable, they were easier to liquidate with apparent justification.

Robert K. Mueller of A. D. Little has pointed out the five stages that the new venture process has gone through since the late 1960s:[11]

1. Euphoria—everybody's doing it
2. Despair—disillusionment sets in
3. Chaos—the specialists and managers pile in to administer artificial respiration
4. Punish—the new venture manager is transferred or fired
5. Reward—a new manager having nothing to do with the failed venture takes over

Task Forces

There are advantages to task forces assembled specifically to deal with new venture management as the need arises. Although such an organization may lack complex planning, formal designations of authority, and long lists of rules, there are many pluses: (1) a consensus as to goals, (2) freedom to act, and (3) a high degree of individual involvement.

It takes some time to get a hastily formed group to operate efficiently, but the new group is highly efficient overall because it is formed only as needed. It is free from preconceived restraints.

The 3M Company sets up new task forces to commercialize new ideas. Allen E. Reynolds of the 3M Company, talking at a joint CMRA-CDA meeting on "Ideamanship" said the real problem is not the idea, but getting the idea developed.[12] It is necessary to minimize the gap between marketing and technical people. 3M sets up two-person entrepreneurial teams of technical and marketing people who can act as if they were actually entering their own small business. They have to answer two questions about the new invention: (1) Can it be made? (technical), (2) can it be sold? (marketing). 3M prefers the small-business entrepreneurial approach with a number of small projects going at once, as against gambling all on one big project with a lot of money poured into it.

Separate Venture Capital Organizations

In 1973 about 30 corporations had formal venture capital organizations (versus 15 in 1970), with about $160 million in capital. The aims of such organizations are both to make profit and to get a window on new technolgies. Such companies include Hercules, Sun Oil, and Johnson & Johnson. But companies that once took such risks and have since given up on the venture capital route include Studebaker-Worthington, Scott Paper, Du Pont, Ford, and General Mills.[13]

The most significant financial bottleneck in this new venture or innovation development process is the limited amount of "seed capital" available for formation and start-up. Only $10 to $30 million a year is spent in the United States for this purpose.[14]

Spin-Offs

In a spin-off, a large company takes a part of itself and gives it independent life, while possibly retaining some degree of interest and control. The spin-off represents a new synthesis in the search for organizational forms suitable for developing new ventures. The spin-off brings together advantages and attributes of small and large companies.

General Electric has successfully used this method to solve a long-standing problem in commercializing research. A large, high-technology company such as

G.E., with a huge research budget, develops many innovations. When one of these innovations appears to have a limited market, the large company has trouble gearing down. General Electric has solved this problem by spin-offs in which G.E. retains a share, as well as giving equity to employees who take off with new ventures.

The several spin-offs are under the umbrella of G.E.'s Technical Ventures Operations, and candidates for spin-offs are evaluated by David J. Ben Daniel, who heads it up. Criteria for spin-offs include:

1. A "niche" market in which the technology will give a unique advantage and also be a wedge for larger markets.
2. A technology that G.E. desires to keep participating in as a "window" in new technology. If G.E. does not have a long-term interest, it will try to sell the technology outright.
3. "Dedicated advocates" among the G.E. employees connected with the new technology who will be capable of taking the risks and responsibilities of a new venture, and who show the other entrepreneurial qualities necessary to make the new venture a success.

Only about one of five proposals examined meets such criteria.

There were six such spin-offs at G.E. in the period from 1969 to 1973. They are in such widely diverse areas as electrical, electronics, educational, oil, metal processing, and filters.

General Electric made its first spin-off venture investment in May 1969, in Applied Materials, which makes equipment for integrated circuit production. It was expected to reach $20 million in sales in 1974. Applied Materials paid back a $1 million loan, and G.E.'s equity rose from $1.5 to $3.2 million by the fall of 1973.[15]

G.E.'s equity in all of its spin-offs was about $4 million at the end of 1973, over three times the book value of G.E.'s original investments at the time of spin-off. However, this did not include G.E.'s losses and research and development expenditures prior to the spin-offs.

An article in *Fortune*[16] tells the story of one of the newer spin-offs, the Nuclepore Corporation, formed in 1972, to exploit a new filter made by punching fine holes in a membrane by means of nuclear radiation. The partners include G.E., a Japanese firm, Nomura Microscience Company, and 14 individuals of whom 6 were with G.E. Such spin-offs offer alternate routes to new venture development.

The handful of scientists and engineers at G.E. combined their limited personal savings with a $550,000 loan from a Japanese firm to back their faith in the new technology they discovered at G.E. General Electric had spent 5 years and $7 million in unsuccessful attempts to make money out of the new filter.

The partners took the following participations:

1. General Electric—41 percent in exchange for licensing the technology and about $380,000 in physical and working capital assets (plus $7 million in lost sunk capital)
2. 15 individual employees—36 percent for a $95,000 investment
3. Nomura—23 percent for a $550,000 investment. The investment is by a venture capital subsidiary of Nomura Securities Company, Japan's largest brokerage house.

The Nuclepore filter was developed in 1965 in the G.E. laboratories. In 1966 the Nuclepore group was put under the Nuclear Energy Division, which was not geared to selling small-ticket specialty items, being used to selling large and expensive nuclear reactors. Five years later, by October 1971, Nuclepore had an investment of over $7 million, sales of only $100,000 a year, and cumulative operating losses of $3 million. The Nuclepore business was too fat—too many people, too many products. It needed a small-company approach. In 1972, G.E.'s Technical Ventures Operation assembled a new management team, slashed the operation to the bone, and arranged for outside investment. The new company was established in December 1972. In its first year of operation the work force was reduced from 36 to 16, sales were doubled from $300,000 to $600,000 a year, and losses were cut from $400,000 to $200,000 a year. At the end of 1973, sales were running at an annual rate of about $1 million a year.

Summary

As has been pointed out in this chapter, the question of how to manage new ventures is far from settled and may never be settled. New changes in the business environment may call for new techniques and procedures. Many elements contribute to the success of new ventures. Management is being increasingly recognized as one of the key elements. Management deals with the control of things and persons. It is a "science" in that it deals with organized knowledge and "art" in that it deals with elements incapable of transfer from one person to another. Management problems are mainly "people problems."

"People problems" are apparent throughout the new venture development process. First, top management must be convinced that new ventures are needed to meet profit and growth goals. Innovation disturbs the order of things. All employees must be convinced to accept innovation. The management of innovation may be more important than innovation itself when the goal is profit. Many innovations are in areas for which the company has not planned. Such unexpected technical innovations need management innovations to exploit the possibilities.

The development of new products seems more successful in small companies than in large organizations. The essence of the small company is the entrepreneur who can put together the other factors of production in such a way to make a profit. Observing small-company success in new business, many large corporations have attempted new venture organizations that simulate the small-company environment within the large corporation. The keys are "small" and "entrepreneurial." Both elements appear very important in commercially successful new ventures. Both the organizations and the people in the organizations have to be entrepreneurially oriented. An understanding of entrepreneurship and a recognition of what constitutes an entrepreneur are vital.

Methods of new venture management have attempted to simulate small company entrepreneurship. Such routes include (1) venture departments, (2) ad hoc task forces, (3) venture subsidiaries, and (4) spin-offs.

One of the more popular routes has been to organize *new venture management departments.* This has been a relatively unsuccessful "fad" route. It was inflexible, tied up in procedures instead of results, and did not tend adequately to motivate and reward the corporate entrepreneurs. In 1970 it was *the way.* Almost one-third of the largest U.S. companies had gone this route. In 1974, few still did. It was too visible, too unwieldly, and too slow to produce results.

Organizing *task forces* when new innovations develop to the point where they can be commercialized has been more successful. It is more flexible and less costly than a formal permanent department. The 3M Company has been most successful in such an approach.

In 1973, about 30 companies had formal *venture capital organizations.* They are aimed at making money and tracking new technologies. The bottleneck is the limited amount of money available to them.

Spin-offs involve a large company giving some degree of independence to a new venture, bringing in other partners and investors. General Electric has been using this route successfully for several years.

It is apparent that the last word in management of new ventures has yet to be said. This is what we know: New ventures should be small, entrepreneurial, and flexible. But we are still not quite sure of how to do this in the framework of the large corporation. If we can learn through past failures, we are advancing. The prime problems are "people problems," and the recognition that this is so is by itself a major advance.

Notes

1. Robert S. Aries and William Copulsky, *The Marketing of Chemical Products* (Brooklyn, N.Y.: R. S. Aries & Associates, 1948).
2. William Copulsky, and Herbert W. McNulty, *Entrepreneurship and the Corporation* (New York: American Management Association, 1974).
3. "A Management Leapfrog at Du Pont," *Business Week,* December 15, 1973, p. 48.

4. Richard M. Nixon, *Message to Congress on Science and Technology,* March 16, 1972.
5. A. D. Little, Inc. and Industrial Research Institute, *Barriers to Innovation in Industry: Opportunities for Public Policy Changes* (Main Report, and Executive Summary) (Washington, D.C.: National Science Foundation, 1973).
6. Mortimer R. Feinberg, "Dealing with People," *Printing Impressions,* November 1973, p. 67.
7. Robert W. Van Camp, "Essential Elements for New Product Success," in *New Product Development,* ed. J. Eastland and J. Tinker (Chicago: American Marketing Association, 1968).
8. Mort D. Margolis, quoted by Richard L. Turmail, "Managers Promote Engineering Direction," *Electronic Design* 26 (December 20, 1973): p. 102.
9. Douglas McGregor, *The Human Side of Enterprise* (New York: McGraw-Hill, 1960).
10. Earl R. Driver, "Organizational Reaction to Change in the Market Environment: A Particular Aspect of Entrepreneurship" (Ph.D. diss., University of Texas, Austin, 1973).
11. Robert K. Mueller, "Venture Vogue: Boneyard or Bonanza," *Industrial Marketing Management,* no. 3 (1973), pp. 47-52.
12. Allen E. Reynolds, quoted in "Brewing New Product Ideas," (a report on the joint meeting of the Chemical Market Research Association and the Commercial Development Association, September 1973, Hershey, Pennsylvania on "Ideamanship–Identifying New Opportunities"), *Chemical Engineering,* November 12, 1973, pp. 108, 110.
13. "Corporate Sugar Daddies," *Forbes,* October 15, 1973, pp. 102. 104.
14. A. D. Little, *Barriers to Innovation in Industry,* (Executive Summary), p. 20.
15. "Corporate Sugar Daddies."
16. Sharon Sabin, "At Nuclepore, They Don't Work for G.E. Anymore," *Fortune,* December 1973, pp. 144-147, 149, 150, 153.

Bibliography

Books

Anyon, G. Jay. *Entrepreneurial Dimensions of Management.* Wynnewood, Pa.: Livingston Publishing Co., 1973.

Argyris, Chris. *Organization & Innovation.* Homewood, Ill.: Irwin and Dorsey, 1965.

Burns, Tom, and Stalker, G. M. *The Management of Innovation.* London: Tavistock Publications, 1961.

Butrick, Frank M. *How to Develop New Products for Sale to Industry.* Englewood Cliffs, N.J.: Prentice-Hall, 1971.

Desrosier, Norman W., and Desrosier, John N. *Economics of New Food Product Development.* Westport, Conn.: AVI Publishing Company, Inc., 1971.

Gerlach, John T., and Wainwright, Charles Anthony. *Successful Management of New Products.* New York: Hastings House, 1968.

Hilton, Peter. *Handbook of New Product Development.* Englewood Cliffs, N.J.: Prentice-Hall, 1961.

Hopkins, David S. *Options in New Product Organization,* Conference Board Report no. 613. New York: The Conference Board, 1974.

Hlavacek, J. D. "A Comparative Empirical Analysis of Managing Product Innovations" (Ph.D. diss., University of Illinois at Urbana-Champaign, 1972), Ann Arbor, Mich.: Xerox Microfilms, 1972.

Karger, D. W., and Murdick, Robert G. *New Product Venture Management.* New York: Gordon and Breach, 1972.

Marting, Elizabeth, ed. *New Products New Profits.* New York: American Management Association, 1964.

Skinner, R. N. *Launching New Products in Competitive Markets.* London: Cassell/Associated Business Programmers, 1972.

Slocum, Donald H. *New Venture Methodology.* New York: American Management Association, 1972.

Stefflre, Volney. *New Products and New Enterprises* (privately printed). Irvine, Calif.: University of California, 1971.

Warshaw, Martin R., and Murphy, Gevara P., eds. *New Product Planning for Changing Markets,* Michigan Business Papers no. 47. Ann Arbor, Mich.: The University of Michigan, 1968.

Wills, Gordon, Hayhurst, Roy, and Midgley, David, eds. *Creating and Marketing New Products.* London: Crosby Lockwood Staples, 1973.

Articles and Talks

Bruck, Donald B., and Wilensky, Samuel, "Start Your Own Business—But Look Before You Leap." *Electronic Design,* June 7, 1974, pp. 142-146.

Chakraborti, Alok K. "The Role of Champion in Product Innovation." *California Management Review,* Winter 1974, pp. 58-62.

Daniel, David Ben. "General Electric's Severed Approach." *1974 Combined Proceedings,* Series no. 36, American Marketing Association, New York, 1973, pp. 478-480.

Davidson, Barbara. "Two Guys Who Made It With Their Umbroller." *New York Times Magazine,* January 7, 1973.

"Developmental Marketing: The Rebirth of the Entrepreneur." *Steel,* September 1963, pp. 50-63.

Fitzell, Philip B. "Meet The Man Who Sold The First McDonald's Hamburger." *Restaurant Business,* November 1974, pp. 81-85, 140.

Gumpert, David. "Playing It Safe, Venture-Capital Firms Fearing a Cash Bind Are Less Venturesome." *Wall Street Journal,* December 4, 1974, pp. 1, 32.

Hanan, Mark. "Corporate Growth Through Venture Management." *Harvard Business Review,* January-February 1969, pp. 43-61.

Herzog, R. H., and Heltzer, Harry. "Minnesota Mining & Manufacturing Company." Remarks to Houston Society of Financial Analysts, April 4, 1973.

Hill, Richard M., and Hlavacek, James D. "The Venture Team: A New Concept in Marketing Organization." *Journal of Marketing,* March 1972.

Hlavacek, James D. "Alternatives in General Management," in *1974 Combined Proceedings,* Series no. 36, American Marketing Association, New York, 1975, pp. 481-482.

Hlavacek, James D., and Thompson, Victor A. "Bureaucracy and New Product Innovation." *Academy of Management Journal,* 1973.

"How G.E. Keeps Ideas Alive." *Business Week,* June 17, 1972, pp. 53-58.

"An Interview with Laser Inventor Dr. Theodore H. Maiman." *Forbes,* September 1, 1968, pp. 42-46.

Klein, Frederick C. "Still Getting Started, Budding Entrepreneur, A Year Later, Finds it Difficult to Run a Firm." *Wall Street Journal,* December 5, 1975, pp. 1, 25.

Peterson, R. W. "New Venture Management in a Large Company." *Harvard Business Review,* May-June 1967, pp. 68-76.

Roberts, Edward B. "A Basic Study of Innovators; How to Keep and Capitalize on Their Talents." *Research Management,* July 1968, pp. 249-266.

"Spinoff Mechanism Offers New Ventures More Chance to Survive, Make Profits." *Marketing News,* September 26, 1975, pp. 9, 12.

Vesper, Karl H., and Holmdahl, Thomas G. "How Venture Management Fares in Innovative Companies." *Research Management,* May 1973.

Westfall, Steven L. "Stimulating Corporate Entrepreneurship in U.S. Industry." *Academy of Management Journal,* June 1969, pp. 235-246.

6
INPUTS TO COMMERCIAL DEVELOPMENT

Newman H. Giragosian
Delphi Marketing Services, Inc.
New York, New York

Joseph A. Vona
Celanese Chemical Company, Inc.
Summit, New Jersey

Edwin S. Michaels
Stauffer Chemical Company
Westport, Connecticut

John C. Trebellas
Celanese Chemical Company, Inc.
Corpus Christi, Texas

Successful commercial development requires the contributions of many company functions. This chapter considers the contributions of three groups that must work together to supply the inputs for successful commercial development.

This chapter will explore the responsibilities of these various groups and the functions that each carries out. For instance, under marketing research, the following questions are discussed:

1. To what depth should marketing research work be carried out?
2. What is the responsibility of marketing research?
3. How does marketing research contribute to successful commercial development?

Similarly, the contributions of the market development and technical service functions will be explored.

Marketing Research

If one were to ask a number of different users of marketing research what the term meant to them, one would probably get a number of different answers. Many definitions of this function have been offered in the past. This has been a reflection of the rapid changes occurring in the chemical industry, which were described briefly in the preceding chapter and will be examined in more detail in subsequent chapters.

In a certain sense, it can be stated that marketing research has been practiced since the beginning of the chemical industry as we know it, even though the function itself was not formally segregated. As marketing research evolved into a separate function, it was restricted to the determination and evaluation of all facts concerning chemical markets. Today, however, it is much more than merely a fact-finding and analysis function. Not only does the marketing reseracher present the findings, but he or she also takes an active role in guiding the company toward making the best decisions consistent with the parameters of the problem to be solved.

In essence, marketing research is a problem-solving activity. The meaning of marketing research can be expressed in a more detailed way. It is directed toward determining the need, acceptance, potential volume, expected competition, possible selling price, most suitable package, use patterns and future trends for already established products and for proposed new products. It means providing data on the sales and potential profits of a given product. Marketing research embraces surveys of the location of the market and specifies acceptable qualities for possible new products.

Marketing research is many things, and successful marketing research has enjoyed such a reputation that it has often been called on to do things that it cannot do. What are they? First, marketing research cannot substitute for mature judgment. This does not mean that marketing research should not recommend courses of action. As a matter of fact, it should and must, if it is to survive in a competitive world. However, marketing research is not soothsaying; there is no crystal ball for the marketing research practitioner to use. This is not to say that marketing research has no predictive value. However, it cannot obtain reliable answers to hypothetical questions.

Marketing research cannot always prove who is right. Too often marketing research directors are asked to prove that someone is right and somebody else wrong. In a clear question of fact, there may be times when such proof can be obtained. Marketing research, should not, however, be asked to perform studies after the fact.

Marketing research is not a question-answer box. Too many senior management personnel still tend to consider it as such: when they tap the box to ask a specific question, out pops an answer. Marketing research, if it is to make a real contribution, must do a truly creative job.

Classification of Major Research Areas and Their Application to Marketing Problems

Established Products for Established Markets

This has traditionally been the area in which marketing reserach has been most used. At the time the book *Chemical Market Research in Practice* was published, emphasis was devoted to this type of marketing research. The prime function of the marketing research department at that time was to determine supply and demand for established products. In a typical demand analysis, the major end uses were determined along with the current and future uses for these materials and explanations of these uses. In addition, the effects of competitive materials on present and future consumption were anlayzed.

The following factors have become part of the marketing research determination of markets for established products.

Marketing Characteristics

Here an evaluation of the major customers for the established product is made. The dominant suppliers for specific end uses are determined, and the importance of specific factors, such as technical service requirements, price, and company relationships, in determining the choice of suppliers are considered. Also, the importance of distributors versus direct selling is evaluated, and specific aggreements or tie-ins between producers and consumers are analyzed, such as toll agreements, long-term contracts, and specific quality requirements. The marketing practices of specific producers are also determined.

Packaging and Distribution

The marketing researcher must determine the major preferences for the mode of shipping of the product, that is, barge, tank wagon or tank car, for example. For instance, are there any specific shipping arrangements, such as the consumer's use of his own trucks? In addition, freight rates for major consuming areas should be determined, as well as the grades of the product that comprise the dominant share of the major markets. Another question that should be answered is the location of the important storage points.

Specifications

Current specifications of present producers are obtained, as well as specifications by grade of material, specific quality requirements, and any anticipated changes in specifications. Technical service requirements are also thoroughly analyzed. Such factors as the number of times a technical service representative calls on the customer are important to assist in the determination of the cost of technical service for the proposed venture. The types of problems the customer has, such as grade, application problems, and quality problems, are also important in this determination.

Price

Historical data on prices must be determined not only for the United States but also for other countries. Any evidence of selling below the list price should be discussed. And most important, the future prices of these established products must be analyzed and determined.

Established Products for New Markets

Another important activity of marketing research in recent years has been to find new markets for established products. This again is marketing research on a very sophisticated, difficult level. As discussed in the previous section, marketing research practitioners must be aware of macroeconomic trends that will affect consumer needs. They must then determine whether the existing product for which they are attempting to find new markets can be improved to allow it to compete in these new markets. Also, they must be aware of whether or not the existing product can be priced to capture the new markets that are opening up. Companies, through their marketing research departments, are coming to know the needs of their customers and their customers' customers far better than they have in the past. New information has been developed on the needs and desires of these customers to the tertiary degree. Also, manufacturing costs should be reexamined, and new ways of making the product must be explored. Marketing research has a fundamental responsibility to be aware of trends that will affect markets for its products and to take advantage of broad trends to increase markets wherever possible.

New Products for Established Markets

In developing new products for established markets, a company can, and often does, follow two routes. In the first, a new product is developed, commonly by the company's research laboratories, and the marketing research group is asked to determine what the possible markets are for this product. In the second, marketing research uncovers new product ideas during the course of broad in-depth surveys relating to the chemical needs for broad end-use industries. A typical example of the first method would be the following.

The first step is to determine the end uses to which the characteristics of the products discovered in the laboratories lend themselves. Next, the products currently serving these end uses should be evaluated. The following questions should be investigated.

1. How large are the markets served by possible products and how fast are they growing?
2. What are the price/performance characteristics of the possible competitive products?
3. What are their advantages or disadvantages?
4. Do the possible competitive products have unique characteristics preferred for the particular end use or uses, or are their advantages a function of price?

5. What are the unique advantages of the new product over currently used products?
6. Who are the customers for the competitive products?
7. What are the channels of distribution?
8. Who are the suppliers of competitive products, and how strongly entrenched are they in the end-use areas served by these products?
9. What are the technical service requirements of the possible customers?
10. What are the probable future prices of competitive products?

The second route, and one that is more commonly used, is to learn the needs of consumers of current products. Information of this type is obtained by the marketing research personnel from technical or research directors in consuming companies. Here the marketing reserach practitioner attempts to determine what the consumer wants in the way of new products and to obtain specifications for an "ideal" product. In addition, he attempts to foresee the needs of the consumer that are not even anticipated at present. He also learns what competitive materials now satisfy these needs, at least in part, and then attempts to discover the specifications, advantages, and disadvantages of these materials. Using this information, he is then able to feed back information on consumer needs to the research director of the company to enable him to more fruitfully orient the work of the laboratory research perscnnel. The questions listed above would then have to be answered after research personnel develop appropriate candidate products.

New Products for New Markets

This is probably the most difficult problem and is truly marketing research at a very sophisticated level. Here, judgment and imagination are important. It is not enough to know merely the current needs of consumers; the needs of consuming populations 10, 15, and even 25 years hence should be anticipated.

This is not to imply that the marketing researcher engages in "blue sky" thinking. He must be able, however, to be aware of broad macroeconomic trends and to visualize their impact on the economy and, specifically, on the end-use areas of the consuming industries. He then envisages the kinds of products and services required to satisfy these developing needs. History is replete with examples of the way in which developing macroeconomic trends have affected the end-use consumption of various products. The automobile is often cited as an example, but a more recent one is the development of synthetic fibers that not only serve established markets but have also invaded markets commonly serviced by other materials and have even generated consumer demands that did not previously exist.

Research on Products and Services

Evaluation of Proposed New Products and Services

Concepts of new products arise from a number of different sources:

1. Ideas from the research and development department
2. Ideas from the engineering department
3. Market development (established during field trips on other products)
4. Ideas from the marketing research department
5. Discussions with research directors of companies interviewed
6. Discussions with marketing managers
7. Discussions with purchasing agents
8. Discussions with technical service personnel

It is important that potential products be screened as soon as possible. It is at this point that marketing research makes a great contribution. Marketing research evaluates, in an exploratory fashion, possible competitive products, analyzing the following factors with relation to these products:

1. Advantages of each
2. Disadvantages of each
3. Prices
4. Price/performance characteristics
5. Specifications of an "ideal" product
6. Need for another product in the end-use area contemplated
7. Commercial feasibility of each potential product
8. Exclusion charts

There are many factors to be considered in evaluating new product ideas, and several departments are involved in evaluating them, among the marketing research, market development, manufacturing, engineering, and the economic evaluation department.

Marketing research analyzes the following factors in evaluating new product ideas:

1. What purpose will the product serve? Will it improve our quality, lower our costs, etc?
2. Would the product fit logically into our existing line of products?
3. What is the anticipated market demand? Are the potential applications for the product broad or restricted? Will the product be a large-volume item or a minor item in the line? What are the prospects for the product on a long-range basis?

4. Will it be pössible to sell the product through our sales force and channels of distribution? If not, what will be the cost of promoting and establishing new channels?
5. Can the product be sold, in general, to the same consumers or through the same distribution channels as our existing products?
6. How long would it take competition to copy or imitate the new product?
7. Is the product technically practicable?
8. Can the item be patented?
9. Does the company have the human resources available or reasonably obtainable to develop, manufacture, and introduce the product?
10. What is the probability of successful development and introduction of the product?
11. What are the prospects for the product on a long-range basis?
12. What is the probability of successful development and introduction of the product?

Comparative Studies of Competitive Products

This has become one of the more important functions carried out by marketing research. On a more or less regular basis, products competing with the company's own present or contemplated products are anlayzed by continuing contacts with end users. Factors such as price/performance characteristics of these and competitive products, technical properties, and product quality are analyzed. Another improtant element considered is whether the demand for these and competitive products is elastic (that is, a function of price) or inelastic.

Test Marketing

One of the areas in which chemical marketing research is becoming more actively involved is that of test marketing. Although this is a costly procedure and is subject to bias, because of the possibility that the sites chosen for test marketing may not be representative, the advantages are very obvious. First, it is possible to obtain actual quantitative data that are not available in any other way. Also, consumer reaction under actual marketing conditions can be ascertained and measured.

Areas in which test marketing can be effectively carried out include the determination of advertising effectiveness, the effectiveness of point-of-purchase displays, and the evaluation of consumer attitudes toward new products.

Other advantages to be derived from test marketing are:

1. It permits careful study of consumer reception.
2. It reveals possible product defects.

3. It localizes sales repercussions of any possible major product difficulty.
4. It reveals sales expectations.
5. It enables advertising appeals to be tested and perfected.
6. It permits price experimentation.

Other Areas of Research

Marketing research is increasingly being drawn into the solution of esoteric problems that up to now were considered the province of other disciplines such as distribution research and advertising research. It is extremely difficult to evaluate the possible contribution of marketing research in these areas because of the many variables involved in measuring the impact of changing forces upon the effectiveness of these disciplines. In addition, packaging research, marketing planning, plant location analysis, evaluation of distributer effectiveness, and competitive pricing research are becoming more important. Some limited work has been done in these areas.

In marketing planning, marketing research can be useful in the following ways:

1. Special studies of products, markets and marketing practices that are directly useful in planning strategies
2. The study of new products and product concepts for broadening the company's product line
3. The evaluation of new opportunities for the company by examining entire industries or specific market vulnerability
4. Measuring various aspects of marketing effectiveness.

Chapter 10 will discuss marketing planning in greater detail. As the forces determining the marketing structure of the chemical industry change, marketing research changes accordingly.

Market Development

Following a decision to proceed, it is important to consider the market development function. What is it? What techniques are available and appropriate? Where should market development report? To sales? Research? A general manager? The president? Or to someone else?

Many in corporate management feel that commercial market development is a sort of obscure functional group that just may come up with the answer to their new product or new profit needs. Others may not even know that they have a market development department within their company. They may care less about who does the job and how: "Just produce profits out of all that dough we're pouring into R&D, the quicker the better. Why, the financial analysts were asking me just the other day what new products we were coming out with, and I had to bluff my way through. I want action and I want it now."

Under these circumstances, market development departments are often hastily organized and expected to produce quick, positive results.

Under pressure the department will go ahead full blast and sometimes come through with flying colors, and then again, sometimes not. The chances favor the latter.

To be consistently successful, market development of potentially profitable new products should be done by an individual or a group of people with a clear set of objectives in mind. They should not be encumbered by tasks that do not apply to their objectives. This must be understood by all who are involved, right up to the responsible executive, president, or general manager.

The market development group should have a fair degree of autonomy in their operations, obviously being responsible for their performance to management within the confines of their objectives. If this is not the case, the chances are that objectives will not be met.

If the market development group reports directly to sales, the director of marketing may take full advantage of their technical expertise by having them spend time on technical service problems. However, the president or division manager may be hell-bent on making acquisitions to quicken short-term profit objectives and will ask the market development group to spend time on these activities, which may be justifiable, but it dilutes the group's effectiveness in getting new products to the market.

If all are agreed that new markets for products is the major objective of the group, some strategies should be established. How many potential customers do we talk to? What sort of presentation are we going to make when we get there? Do we (or can we) talk price during the initial visit? Is there money in the research budget to perform additional applications studies? The strategy should involve the action plan by anticipating, at least in common-sense terms, how we plan to accomplish the objectives within the expected time period.

All of that input that went into the decision to go ahead with the market development should be reviewed periodically. Sometimes it is possible to discover that an important piece of information may have been overlooked in the initial decision. Marketing research may have seen that the new product might have a relatively short span of life as a profit producer because of new technology that was not known at the time of the first survey. Also, inflation may dictate that plant capital investment for a commercial facility to be completed 2 to 3 years in the future could knock that estimated return on investment figures into a cocked hat.

An understanding of the business area to be entered with a new product is absolutely crucial. An evaluation of usual business practices in this area may uncover all kinds of wrinkles and deals unknown to the researchers if their company is not currently active in the area.

Research people often may assume that their new invention should capture the major portion of an existing market because it performs better in laboratory tests. It may be that potential customers could care less about quality nuances, being more interested in getting maximum profit from lower-performance products.

The importance of keeping research informed on the field acceptance (or rejection) of a new product cannot be overemphasized. One of the best ways to keep research informed is to take laboratory people out to the field for firsthand exposure to the potential customers and their reactions to the new product. This is far better than relying on them to get their feedbacks from customer call reports, which often contain a predominance of subjective editorialization.

Coordination with the sales department is no less important than communication with research because the new product will eventually be turned over to sales once it has reached commercial stage.

The transition to the regular sales group can be facilitated by keeping them informed of the market development's activities from initial field introduction to product maturity.

Likewise, production people should be advised on the development projects because they must become involved in manufacturing successful new products. In many cases, the pilot and semicommercial quantities made for supplying the new markets will be made by production groups. The earlier they know about the details of the potential quantities and product specifications, the better.

With all of this background in mind, it might be worth considering two actual market development projects with which I am personally familiar—one successful with respect to generating acceptable gross profits and one that "bombed."

The Successful Product

The company in question had considerable background in the technology of flame retardants, considering that the market then was rather embryonic. From the technical side it was basic in many flame-retardant raw materials, that is, phosphorus and chlorine; basic and applied research personnel were knowledgeable and productive, production facilities were adequate for existing fire-retardant compounds. Only the *market* wasn't there.

Enough money had been appropriated and spent in research each year so that management was increasingly concerned about the less than adequate profit returns resulting. Although basic chemicals such as ammonium phosphates were being sold in the boxboard and decorative areas, the market development group was getting tired of saying that the day of more durable, more sophisticated fire retardants was just around the corner. This crystal-balling, naturally, did not help management toward their profit objectives.

Then from the field sales group, the need was defined for a durable, more permanent fire retardant for rigid urethane foam. The potential market was unknown because foamed urethane itself was relatively new to the insulation market. But this market seemed to be large enough to warrant at least a few man-months effort in basic and applied research.

When all agreed that this was an opportunity, research quickly synthesized some 13 compound candidates, all of which were designed to fit the qualification that the product become an integral part of the finished foam and render adequate fire-retardant qualities to it under conditions of manufacture and use.

After screening by applied research, five of the original products were selectively sampled to potential customers in the rigid-foam field. Selection of companies was made on the premise that because the fire retardant supplier was not really in the urethane foam business area and because they did not really have too much product to go around, they had to depend on potential customer feedback to help direct the development program.

Sampling, resampling, and drum quantity shipments took place in succession over a 9-to-12-month period. During this time, one of the original candidates began to give consistently better results in the laboratory and the field tests.

Initial pricing estimates indicated that the product had to be priced relatively high with respect to what the users would pay, or at least said they would pay, in the early stages. This led to sharper questions from the development group with respect to competitive product pricing and effectiveness, because by this time, the market need was becoming apparent to other potential producers of fire retardants.

There were two principal courses of action open: keep the price relatively high, as had first been the strategy, or drop it to more competitive levels and hope to gain a larger volume share to try to make up for the lower per-pound profit. The former course was chosen for two reasons: The total markeplace needs were still relatively unsettled, and competitive products seemed to be doing a satisfactory job in the lower-quality markets. Moreover, were the company to drop its price, competition could probably drop theirs also. In summary, the decision was made to go after the high quality, lower volume segment and watch for opportunities in the larger segment in case competitive products fell on their faces or unexpected economies of manufacture allowed a lower pricing structure for the new product.

As it turned out, competition did not get as big a share as expected, and economies of manufacture simply did not exist for a negative gross profit producer. Production difficulties in the modest-sized plant existed over a 2-year period, finally being fairly well corrected by adding appreciable new investment to eliminate bottlenecks.

Marketing research continued to experience a great deal of uncertainty about rigid foam in general and fire retardants foam in particular. Volume forecasts made 4 to 5 years into the future varied as much as 400 to 500 percent. It was extremely difficult to predict how things were going to happen because the forecasts were all coming from leaders in the foam production field. One of the major uncertainties related to the acceptance of foam in the construction industry. All in all, though there was general agreement that the future looked bright and that some sort of fire-retardant products would have to be used.

As the sales volume for the additive in question increased, the field sales force became more active as they became aware of the potential market aspects and the technical problems involved. Joint follow-up calls by development and sales personnel became more effective, and it was decided that certain field salespeople could follow up some accounts. This allowed development to concentrate more on the "problem accounts."

When it was decided that the product had reached the degree of maturity necessary to be fully accepted by the sales department, it was relatively easy to turn the principal sales responsibility over to them because they had become quite knowledgeable regarding existing and potential customers. Where field activities had been modest, group seminars with field sales personnel were held.

After turnover, the market development and marketing research groups were free to work on successor candidates to protect the commitment to this business area.

The Unsuccessful Product

Back in the days when research was given a free hand to work on projects of their choice, it was not at all unusual to grind out copious quantities of "interesting" compounds, advertise them, and then have someone go find a market for them. Every so often, applications work proceeded to the point where those involved could get excited about the commercial prospects after seeing an in-house lab test or two. Brochures were printed, pilot quantities prepared, sometimes elaborate cost-price estimates were worked up, the potential customer list prepared, and market development people went forth to the field. More than a few times they found a lukewarm acceptance because too many vague assumptions had been made in-house prior to determining any market need. This was especially true where the area involved was new to the company doing the product introductions.

The case in point here involved the introduction of a new surfactant to the dry-cleaning trade. Some marketing research had been done in the field by a company that essentially had little expertise in the end-use area. Initial reports prompted a rather extensive synthesis program to come up with a product having several superior quality aspects compared to the materials that the trade was using.

Although several candidates showed interesting properties in the applications lab, the one chosen for field introduction turned out to be a high-cost, high-priced product. Some rationalization was given to the presumption that it should be billed on the basis of better cost-effectiveness, even though the product would have to sell at between two and three times the cost of the currently used surfactants. Also, some of the performance negatives appeared to be ones that could be "formulated around" by people "knowledgeable in the technical end."

Somehow, details on the product got out to researchers in the field even before market development was ready to begin sampling. This caused delays in responding to sample requests and eventually forced the company to supply samples to people that the development department had not intended to contact initially.

From the start, the field introductions gave some strong indications that formulators were rather well satisfied with their existing, lower-priced commercial surfactants. Nevertheless, they agreed to put the new candidate "through the paces" and report back on the results. It was not too surprising, in view of the rather cool initial receptions, that these results did not come in very promptly. In fact, one customer took almost 2 years to give a meaningful, documented, and predictably negative response.

As the postmortems took shape, it appeared to all concerned that the chances of both technical and commercial success would have been greater if the company had been in the business area to begin with. Lacking this advantage, more in-depth marketing research coupled with a greater commitment to the surfactant area in general could have bettered the odds for success.

In any event, the company in question learned by firsthand experience that, particularly in today's business climate, there are many factors that go into the successful commercial development of new products, not just successful performance in their own laboratory.

The Scope of Technical Service Activities in Commercial Chemical Development

Historical

The concept of a technical service department or a technical service laboratory has evolved with the historical development of modern marketing techniques in commercial chemical development. Before its origin, personnel in either the engineering and manufacturing department or in the research and development laboratory provided a rudimentary form of technical service. But, as the need for a fully staffed technical service laboratory became recognized by management, a separate and distinct organization was established and assigned these responsibilities.

Multiple Functions

Perhaps the best way to understand and appreciate the need for technical service in commercial chemical development is through a demonstration of its multiple functions. To accomplish this, it is best to discuss in some detail the more important areas of technical service responsibility.

Servicing the Customer's Technical Needs

Technical service is an important factor in developing the loyalty of existing customers. This is because it handles the customer's technical problems and complaints and thereby makes his or her work less difficult. In this way, it helps the customer improve his or her performance and profitability.

New customers generally will not buy a product unless they are sure it will serve their requirements. The best way to convince them that a product will perform is to offer them complete technical data and back it up with the required amount of technical service. By assuring customers that their problems are yours, it is possible to differentiate your product when price and quality do not differ substantially from one brand to another. In essence, this reflects the use of technical service as a nonprice competitive sales tool.

In all respects, technical service symbolizes the goodwill of the seller to technically back his or her product. If a supplier's service activities are valued by the customer's production and technical personnel, they will, in turn, favorably influence their purchasing department.

Develop New Applications for Existing Products

The technical service department also has the major responsibility for establishment of new uses for existing products. This often requires the development of new formulations or systems incorporating the products and tailoring variations for specific requirements. Generally, the objective is to expand and broaden the market for existing products—particularly into specialized end uses that can support the maximum profit margins. Because of the rapid technological changes occurring in the chemical industry and the increasing expense of research, most application research programs are being limited to those of shorter range and higher relevancy.

Applications research is sometimes considered a form of aggressive or offensive technical service as distinct from defensive technical service, which is done primarily to satisfy a specific technical problem faced by customers.

Assist New Product Development

The technial service staff can be an important aid to new product planning and development. It can help in identifying market needs that suggest new products.

Because technical service is highly concerned with the customer's requirements and objectives, it will often find a product essential to solving the customer's problem. The success of this kind of effort is critically important to many customers, particularly those less technically sophisticated. It is also a fruitful source of many new product ideas.

New product involvement is also helpful in increasing the effectiveness of the technical service staff. By being a part of new product development, the technical service department is able to generate the bulk of information that will inevitably be needed to answer questions submitted by the customer, the marketing department, and the corporate planning staff.

Involvement in new product development is very important to the technical service department as well. Through this function, technical service is able to offer its invaluable input into new venture analysis and commercial development efforts.

Services to the Sales Department

Technical service helps protect sales and profits. By providing effective technical service, it helps offset price competition.

Services to the sales department are provided in numerous ways. The more common services offered are the giving of technical seminars, making product demonstrations, or the writing and dissemination of technical bulletins. The technical service staff also assists sales by making customer calls with sales personnel and by giving them technical training and orientation courses.

The services provided to the sales department afford the opportunity for improving sales performance by relieving the sales staff of technical duties. As a result, the sales staff is better able to spend more time on cultivating the customer and on the development of sales techniques.

Keeping Abreast of Competitive Products and New Technological Developments

Another important function of the technical service staff is to examine and evaluate existing competitive technical literature and compare it to internal data. Any technological trends or changes in the industry should be noted. Both the marketing and research departments are immediately advised of changes noted in existing products or the introduction of a competitive new product.

Provide Technical Liaison Between Marketing Research and Technical Consultants

Because of its dual role as a marketing and research organization, the technical service staff is required to bridge the gap between the fundamental research and the marketing departments. In this function, technical service must often reconcile conflicts in judging product potential, performance, and technical capabilities. This is a key responsibility that serves the broad interests of the corporation.

Technical service personnel acquire extensive contacts within industrial laboratories, research foundations, trade associations, government laboratories, and with key consultants. These contacts are invaluable when information and services are needed outside the scope of the technical organization. Active participation and memberships in technical societies and trade associations are also an essential activity and an important source of information.

Organization

There are several options for the organizational positioning of the technical service department. It is usually placed in the corporate organizational structure as the technical staff of the marketing department. Many exceptions to this structure do exist. Placement of the technical service function is occasionally affected by the managerial needs of the business at a particular time or by the abilities and background of the personnel on the technical service staff. To fill a particular need, the technical service department may be a distinct part of the research and development organization or a separate staff organization within a commercial development department. Placement is usually influenced by what management feels is the most important of the numerous functions the technical service department performs.

Almost invariably because of its specialized responsibilities, the technical service function is a staff activity regardless of which line organization it reports to. It should be noted, however, that the abilities of the technical service function are influenced by its line organization. Placing technical service with either the research or marketing organizations presents different environments. Placing the technical service people within a typical research organization will enhance their technical abilities but could result in a loss of their understanding of the realities of the marketplace and its needs. In other words, with the technical service group in a marketing organization, there is a closer contact with the customer and a better realization of the requirements for commercial product development. This approach seems preferred it the marketing and sales departments can be kept from exerting the pressure of short-term sales problems on the technical service personnel and thereby destroy their objectivity.

The ideal organizational placement for the technical service function is obviously quite difficult to generalize. Corporate marketing philosophy in each company will dictate the structure. Whatever the position established for technical service, it is imperative that complete freedom of contact with other departments be allowed to assure its maximum effectiveness.

Personnel

In order to be totally successful in its functions, the technical service department must be staffed with professionals of recognized technical competence in the fields of industry the company services.

The type of individual best suited for the technical service staff is rather unique. He must be a cross between an effective salesman and a creative researcher. He should have a natural inquisitiveness that will keep him aware of industry conditions that may benefit his company. He should not be prone to make rash or hasty decisions when under stress. When stumped for an answer to a question, he must not easily get discouraged although he is constantly faced with technical problems and customer complaints. He must have attitudes and motivations that are attuned to swiftly changing business environment. Finally, and most important, he should be able to communicate effectively with all kinds of people—management, sales, plant, technical and so on—in both written and oral form.

Technical Service Today

Providing adequate technical service is very expensive in terms of money, material, and time. The chemical industry has been traditionally an industry of high capital intensity and rapid overexpansions. This has invariably resulted in many cases of costly overcapacity with prolonged periods of highly competitive selling and low profit operation. During these profit squeezes, sales may increase but are usually accompanied by a decline in return on investment. At such times, technical service is an expensive but entirely essential function for maintaining customer goodwill and loyalty.

In more recent times, the chemical industry has drastically changed. The industry is now faced with severe energy and hydrocarbon raw material shortages at a time of increasing market demand. As a result, many firms have been forced into product reevaluations. They have called upon the technical service laboratory to help in reformulating products to more economical, less energy-sensitive products. In a climate of raw material shortages, some aspects of technical service may tend to be restricted. For example, customers who are able to receive material are usually so pleased that they hesitate and are reluctant to request technical service from their suppliers.

Technical service itself is being reevaluated because of rising costs. Many large corporations have instituted new policies limiting technical service to those functions considered essential to the business. The technical service manager has been restircted in budget and asked to ensure that his or her department is contributing to the company profits. Like others, he or she must assure management that their investment in technical service is worthwhile. To do this, all technical service functions and applications research programs must be continually evaluated to see that they are on goal. A change in probability of success or market conditions may require a reevaluation and perhaps elimination of a project to make room for one more highly rated.

Although it is considered an essential sales tool, even the customer service function must be placed under cost controls. Different customer needs will dictate a priority of themselves as being critical or not. It is important that abuses of too liberal a service policy be avoided. For example, it is not unusual for some customers to demand services out of proportion to those justified by the profits they generate. When this happens, some restrictions must be imposed to prorate services to the profit potential of the customers. There are many ways in which customer services can be restricted, and each organization must establish some controls to ensure their profitability, while being equitable to their customers.

In conclusions, although restricted in many of its functions in today's market, nevertheless because of its strong contribution to expanding sales and new product market development, the technical service function will expand with the chemical industry. After considering the numerous responsibilities and uses of the technical service departments, it is easy to understand why it is said that as long as chemicals are offered for sale, a technical service staff will be needed.

Bibliography

Adler, Max K. *Modern Market Research.* New York: Philosophical Library, 1957.

Alderson, Wroe, and Shapiro, Stanley J. *Marketing and the Computer.* Englewood Cliffs, N.J.: Prentice-Hall, 1963.

Alevizos, John P. *Marketing Research; Applications, Procedures and Cases.* Englewood Cliffs, N.J.: Prentice-Hall, 1959.

Alexander, R. S., Cross, S. C., and Cunningham, R. M. *Industrial Marketing.* Homewood, Ill.: Richard D. Irwin, Inc., 1964.

Balthasar, H. U., Boschi, R. A. A., and Menke, M. M: "Calling the Shots in R&D." *Harvard Business Review,* May-June 1978, pp. 151-160.

Blankenship, A. B., and Doyle, J. B. *Marketing Research Management.* New York: American Management Association, 1965.

Bogart, Leo. "How to Get More out of Marketing Research." *Harvard Business Review,* January-February 1956.

Borden, Neil H., and Marshall, Martin V. *Advertising Management: Text and Cases.* Homewood, Ill: Richard D. Irwin, Inc., 1959.

Boyd, Harper W., and Westfall, Ralph. *Marketing Research: Test and Cases.* Homewood, Ill.: Richard D. Irwin, Inc., 1964.

Bradford, Ernest S. *Marketing Research: How to Analyze Products, Markets and Methods of Distribution.* New York: McGraw-Hill Book Co., 1951.

Brenner, Henry, ed. *Marketing Research Pays Off.* Pleasantville, N.Y.: Printers' Ink Books, 1955.

Brion, John M. *Corporate Marketing Planning.* New York: John Wiley & Sons, 1967.

Bross, I. D. J. *Design for Decision.* New York: The Macmillan Company, 1953.

Brown, L. E. *Marketing and Distribution Research.* New York: The Ronald Press Company, 1955.

Butrick, Frank M. *How to Develop New Products for Sale to Industry.* Englewood Cliffs, N.J.: Prentice-Hall, Inc., 1971.

Buzzell, Robert D. *Mathematical Models and Marketing Management.* Division of Research, Graduate School of Business Administration, Harvard University, Boston, Mass., 1964.

Chaddock, Richard E., ed. *Chemical Market Research in Practice.* New York: Reinhold Publishing Corp., 1956.

Clewett, Richard M. *Marketing Channels.* Homewood, Ill.: Richard D. Irwin, Inc., 1954.

Conrad, E., and Erikson, R. *Marketing Research Techniques.* New York: American Management Association, 1966.

Corley, H. M., ed. *Successful Commercial Chemical Development.* New York: John Wiley & Sons, 1954.

Crisp, Richard D. *Marketing Research.* New York: McGraw-Hill Book Co., 1957.

Dash, Joseph F., Berenson, Conrad. "Techniques In Marketing Research (Keeping Informed." *Harvard Business Review,* November-December 1969, p. 14.

Delens, A. H. R. *Principles of Market Research.* London: Crosby Lockwood, 1950.

Developing a Product Strategy. (Chapter XI: Testing for Market Reaction), Management Report No. 39, New York: American Management Association, 1959.

Enright, Ernest J. "Market Testing." *Harvard Business Review,* September-October 1958.

Establishing a New-Product Program. Management Report No. 8, New York: American Management Association, 1958.

Ferber, Robert. *Statistical Techniques in Market Research.* New York: McGraw-Hill Book Co., 1949.

Ferber, Robert, et al. *Marketing Research.* New York: The Ronald Press Company, 1964.

Ferber, Robert, and Verdoorn, P. J. *Research Methods in Economics and Business.* New York: The Macmillan Company, 1962.

Fox, Willard M. *How to Use Market Research for Profit.* Englewood Cliffs, N.J.: Prentice-Hall, 1950.

Frank, Nathalie D. *Market Analysis: A Handbook of Current Data Sources.* New York: The Scarecrow Press, 1964.

"Generating New Product Ideas." CBR 546, 1972.

Giragosian, N. H., ed. *Chemical Marketing Research.* New York: Reinhold Publishing Corporation, 1967.

Gisser, P. "Taking the Chances Out of Product Introductions: Using the PERT Technique." *Industrial Marketing,* May 1965.

Gold, Jack A. "Testing Test Market Predictions." *Journal of Marketing Research,* August 1964.

Gould, Douglas P. *Marketing for Profit.* New York: Reinhold Publishing Corp., 1961.

Gunther, Edgar, and Goldstein, Frederick A. *Current Sources of Marketing Information: A Bibliography of Primary Marketing Data.* Chicago, Ill.: American Marketing Association, 1960.

Haas, Raymond M. *Long-Range New Product Planning in Business.* University of West Virginia, Morgantown, W.V., 1965.

Hanan, Mack. "Concept Advertising." Management Bulletin No. 37, New York: American Management Association, 1963.

Harder, Virgil E., and Lindell, Frank R. "Using PERT in Marketing Research." *Business Horizons,* Spring 1966.

Hilton, Peter. *Handbook of New Product Development.* Englewood Cliffs, N.J.: Prentice-Hall, 1961.

Hobart, D. M., ed. *Marketing Research Practice.* New York: The Ronald Press Company, 1950.

Holmes, P. M. *Marketing Research: Principles and Readings.* Cincinnati, Ohio: South-West Publishing Company, 1960.

Hummel, Francis E. *Market and Sales Potential.* New York: The Ronald Press Company, 1961.

Kolliner, S. A., Jr. "Measuring Industrial Advertising," *Journal of Marketing,* October 1965.

Lorie, J. H., and Roberts, H. V. *Basic Methods of Marketing Research.* New York: McGraw-Hill Book Co., 1951.

Marketing Research: A Guide for Capital Goods. Chicago, Ill.: Council for Technological Advancement, 1955.

Marketing Research in Action. Studies in Business Policy No. 84, New York: National Industrial Conference Board, 1957.

Marting, Elizabeth. "Reporting Sales Data Effectively." New York: American Management Association, 1959.

Marting, Elizabeth, ed. *New Products/New Profits.* New York: American Management Association, 1964.

"New Product Development." Conference Board Record, September 1976, pp. 25-28.

Options in New-Product Organization. CBR 613, 1974.

Pessemier, Edgar A. *New-Product Decisions.* New York: McGraw-Hill Book Co., 1966.

Redmayne, P., and Weeks, H. *Market Research.* London: Butterworths, 1951.

Scheier, Fred T. *Modern Marketing Research.* Belmont, Calif.: Wadsworth Publishing Company, 1963.

Schuckman, Abe, and Berg, Thomas L. *Product Strategy and Management.* New York: Holt, Rinehart & Winston, 1963.

Slocum, Donald H. *New Venture Methodology.* New York: American Management Association, 1972.

"Sources of Corporate Growth," Experiences in Marketing Management, 24, 1970, pp. 7-9 and 19.

Talley, Walter J., Jr. *The Profitable Product: Its Planning, Launching and Management.* Englewood Cliffs, N.J.: Prentice-Hall, 1965.

"The Role and Organization of Marketing Research," Experiences in Marketing Management, 20, 1969.

Twedt, Dik Warren. "What Is the 'Return on Investment' in Marketing Research?," *Journal of Marketing,* January 1966.

Vizza, Robert F. "Measuring the Value of the Field Saels Force." a research report published by the Sales Executives Club of New York, 1964.

Wales, Hugh G., and Ferber, Robert. "A Basic Bliography on Marketing Research." Chicago, Ill.: American Marketing Association, 1963.

Ward, Richard J. "Forecasting Product Consumption with Computers." *The Engineering Economist,* Spring 1965.

Wasson, Chester. *Research Analysis for Marketing Decisions.* New York: Appleton-Century-Crofts, 1965.

Weir, Walter. *On the Writing of Advertising.* New York: McGraw-Hill Book Co., 1960.

Westfall, Ralph, et al. "Problem Definition." Chicago, Ill.: American Marketing Association, 1958.

"Westinghouse: Tryout in Philadelphia. Is It Possible to Test Market an Industrial Product Accurately and Effectively?." *Sales Management,* Nov. 5, 1965.

Weston, John F. "Defense Space Market Research." Massachusetts Institute of Technology, Cambridge, Mass., 1964.

Whitney, Elwood. *Symbology.* New York: Hasting House Publishers, 1960.

Wilson, Allan R. "Qualitative Market Research." *Harvard Business Review,* January-February 1952.

Wong, Yung. "Critical Path Analysis for Product Planning." *Journal of Marketing,* October 1964.

Zarecor, William D. "High-Technology Product Planning." *Harvard Business Review,* November-December 1975, p. 108.

Zober, Martin. *Marketing Management.* (Chapter 9: "Operations Research and Electronic Data Processing in Marketing Management), New York: John Wiley & Sons, 1964.

PART II

MANAGING THE FUNCTION

7
UNDERSTANDING THE ENVIRONMENT OF NEW BUSINESS VENTURES

George F. Sharrard
Kennecott Copper Corporation
Lexington, Massachusetts

In the world of business today, the objective is more of everything that is good. With today's escalating rate of change, a business organization must continuously seek out new business to replace old business lost because of obsolescence.[1] Only in this fashion can the management of a business assure the owners of profitable continuance of the business. Further, today's financial climate places a premium on those business organizations who, in addition to maintaining profitable operation, also expand their businesses while still maintaining profits. Simply maintaining position is no longer acceptable—today, growth is required. The three key elements of purpose for any business today are profits, survival, and growth.

As the rate of change and the demand for growth increased, so too did the sophistication of commercially developing new business opportunities. The search for new opportunities intensified, until today business finds itself with more opportunities than it has resources. Having identified these opportunities, the next step is that of evaluating them and selecting those that meet the company's objective. Opportunity has to be carefully balanced against risk.

The selection of new business opportunities is the basis for future corporate performance. Good evaluation weeds out the poorer opportunities, highlights the better opportunities, and improves the overall odds of successfully developing future profitable businesses.

The technique of evaluation has evolved into a sophisticated operation. For this reason, consciously and by intent, there are no checklists or empirical procedures, no computer programs, no formulae, ratios, indices, or other magic numbers in this chapter that would guarantee evaluation and selection success. There are no shortcuts to sound evaluations. Every business has its own set of strengths and weaknesses and must tailor its evaluation technique to maximize results. This chapter will endeavor to organize today's thinking on evaluation into an orderly procedure founded on basic concepts applicable to most situations. Using this procedure, a business can develop its own special evaluation system serving its specific needs. But because evaluation techniques also change rapidly, the reader is cautioned against blind use of any set system. Rather, it is suggested that basic concepts be examined occasionally to detect, adjust to, and stay abreast of, change as it occurs.[2]

Business opportunities, like people, come in many shapes and sizes. Each one has its own particular set of characteristics making it different from all others. But they do lend themselves to broad classification. In general, opportunities in business fall into one or a combination of four categories. They involve singly or together *products, processes, systems,* and *services.*

Any given opportunity can involve the domestic or international world of trade. It can arise from internal activities of the corporation or may be recognized in the external environment composed of other business firms, universities, consultants, institutes and foundations, government facilities, private inventors, and others.

A study published several years ago[3] reports the methods used by about 250 U.S. corporations to secure growth in 1972. The sample covered a broad range of industry throughout the United States. The findings, although qualitative in nature, are quite instructive in developing a profile of U.S. business. Of those corporations reporting:

74 percent operated with formal plans for growth

74 percent budgeted for "basic and/or applied" research

56 percent engaged in one or more acquisitions (see Table 1)

33 percent engaged in one or more divestures (see Table 2)

30 percent purchased and sold technology in the form of patents or "know-how"

The volume and complexity of activity indicated by the data in Tables 1 and 2 very well illustrate the necessity for sound and effective evaluation techniques.

Whether operating today's business or planning tomorrow's business, management must take cognizance of those forces shaping and affecting the future of their individual business. Some of these forces are external to the business and relatively uncontrollable. Others involve internal matters and are mostly controllable. The latter are great in number and highly complex.

All these forces change with time, and it is important to evaluate them in the appropriate time frame. Time spans used for planning growth vary depending on the nature of the business; for example, this span may be a few months in

TABLE 1

Acquisitions in 1972 (141 Companies Making Acquisitions)

	Number		
Nature of Acquisition	Domestic	Foreign	Total
Individual products	8	3	11
Complete product lines	17	2	19
Subsidiaries	13	6	19
Divisions	23	4	27
Corporate entities	83	43	126
Total reported	144	58	202

TABLE 2

Divestitures in 1972 (83 Companies Making Divestitures)

Nature of Divesture	Percent of Total
Individual products	18
Complete product lines	27
Subsidiaries	28
Divisions	27

the fashion world and generations in the timber business. Industrial businesses frequently use a 20-year time frame. However, the selection of a time frame for evaluating each business opportunity should be based on the nature of that particular opportunity, not on the basis of conforming to some standard practice.

Evaluation and Selection Procedure

Experience has shown the advantage of systematically organizing these forces as an aid in ensuring that none are overlooked during evaluation. For present purposes these forces theoretically can be organized into three broad classifications: long-term trends, external factors, and internal factors.

Long-Term Trends of Western Culture

Because the time span chosen for this discussion is 20 years it is very important to consider what the world will be like 20 years from now. Many have tried to prophesy the future and from their endeavors has arisen a developing science of "futurology." Exemplified by ongoing work at the Hudson Institute, the first class of forces to be considered in evaluating business opportunities are what Herman Kahn, Director of the Institute, calls the long-term trends of Western culture. These are shown in Table 3 and form the foundation for all evaluations except those covering very short periods of time. They change slowly and, once understood, supply continuing guidance. These factors merit more detailed explanation than can be given here, and the reader is referred to the source publications.[4]

As Kahn points out, trends such as those in Table 3 depend on large, aggregative phenomena and are often more amenable to long-range planning than those that depend on unique circumstances or a special sequence of events.

TABLE 3

The Long-Term Multifold Trend of Western Culture

1. Increasingly sensate [empirical, this-wordly, secular, humanistic, pragmatic, manipulative, explicitly rational, utilitarian, contractual, epicurean, hedonistic, etc.] cultures
2. Bourgeois, bureaucratic, and meritocratic elites
3. Centralization and concentration of economic and political power
4. Accumulation of scientific and technical knowledge
5. Institutionalization of technological change, especially research, development, innovation, and diffusion
6. Increasing military capability
7. Westernization, modernization, and industrialization
8. Increasing affluence and [recently] leisure
9. Population growth
10. Urbanization, recently suburbanization and "urban sprawl"–soon the growth of megalopoli
11. Decreasing importance of primary and [recently] secondary and tertiary occupations; increasing importance of tertiary and [recently] quarternary occupations
12. Increasing literacy and education and [recently] the "knowledge industry" and increasing role of intellectuals
13. Innovative and manipulative social engineering; i.e., rationality increasingly applied to social, political, cultural, and economic worlds as well as to shaping and exploiting the material world–increasing problem of ritualistic, incomplete, or pseudo rationality
14. Increasing universality of the multifold trend
15. Increasing tempo of change in all the above

However, forces in the latter categories of unique circumstances and special sequences of events have to be evaluated by businesses. Establishing a framework within which these fall is helpful. A suggested system for classifying such forces is discussed in the following section.

External Factors

Forces largely outside the control of any single company can be grouped into four classes: economic, social, political, and technical. In each class there are the normal or logical alternatives to be considered as well as a "suprise situation," the latter unpredictable but many times of great significance. In discussing the four classes, examples will be given. Although a force is placed in one particular class, many could be placed in other classes. This is not important. What is important is to identify all the forces and to evaluate their effects on the business opportunity under consideration. For any particular business opportunity, one force or combination of forces will be most important. This will vary by the nature of the opportunity. Thoroughness in assessing each force and its impact cannot be overstressed. For this reason no importance should be attached to the *order* of classes.

Economic

Business cycles with their booms and recessions typify economic forces. They are international as well as domestic in nature. Their effects are well known and should be taken into consideration during evaluation, not left in a statement such as "barring recessions, and other economic disruptions." Some products are more recession-proof than others. Since World War II, nondurables have fared better than durables in U.S. business cycle adjustments. Service businesses based on repairs may run counter to business cycles as consumers repair the old and defer purchase of the new. The AMP Corporation, specializing in electrical connectors, has with two exceptions increased sales every year since 1950. Sales fell twice, one year coincident with the 1957-1958 recession and in 1975. AMP weathered the other three business adjustments, increasing sales each year despite weaknesses in the economy. Determine *how* business cycles will affect the business opportunity being evaluated and set forth the effects clearly.

Taxes are powerful economic forces. In the United States, for example, the federal government allows extractive industries a reduction in federal income taxes for depletion of their reserves. In 1969 Congress revised this depletion allowance, reducing the allowance on crude oil about 15 cents per barrel and increasing the depletion allowance on oil from shale about 15 cents per barrel. The effect on the petroleum industry was immediately to increase their costs, but the most interesting long-term effect may have been on the efforts of a totally different industry, one endeavoring to develop liquid fuel from shale.

After intensive scientific research over many years, the ultimate goal of producing cost-competitive oil from shale still eluded the shale oil interests. At the stroke of a pen, the signing of the law reducing depletion allowances on crude oil and increasing those on shale oil closed a large part of the gap science had been unable to close. Liquid fuel from shale is that much closer to economic feasibility because of tax legislation.

Other economic forces exerting pressures on business are *tariffs, international trade negotiations, monetary policy,* and so forth. Serious reflection will identify others and highlight their ever-changing importance in evaluating the risks and opportunities of new business.

Although this class of forces is perhaps less prone to contain "surprise situations," certain economic events have contained a high element of surprise as evidenced by the government's freezing of wages and prices in 1971. This could not have been predicted 20 years (or even 5 years) previously. Such surprise situations stress the need for continual surveillance and selection of business opportunities having maximum flexibility consistent with meeting growth objectives.

Social

Demographic forces such as changes in population number, age, sex, literacy, affluence, and migration receive intensive study by many government agencies. Statistics are readily available and are generally quite accurate. The effects on markets and their location and size are regularly evaluated by businesses.

Consumer businesses study life-styles, changing consumer likes and dislikes, and motivation stimuli. Industrial businesses are usually less directly involved. However, synthetic fiber manufacturers, for example, study the consumer in depth and evaluate their industrial business opportunities accordingly.

It can be argued that *wars* do not belong under a social classification. But for the purpose of this discussion they have been so classified. There have been four major wars in this century in which the United States has been involved. In retrospect, wars have spawned important postwar businesses. For example, World War II ended with large-capacity aluminum plants and insufficient uses for the aluminum. The aluminum industry seized the opportunity and, through aggressive commercial development, set a growth rate far in excess of normal industrial performance.

The chemical industry was incubated in World War I. Deprived of imported chemicals because of the war, the fledgling U.S. chemical industry seized this opportunity to prove itself and set a high rate of growth for over 30 years. The entire history of the aerospace industry has been written around war. The helicopter, so often used in Vietnam, may well be that war's contribution.

Pressure groups have assumed greater importance in the last 10 years. Minority groups based on race, religion, or sex; environmentalist campaigning for a more protected environment and better life-style (judged by their standards); hippies, the drug culture, communes—all are forces effecting business opportunities. Their effects on business opportunities have received such widespread coverage in the press that no examples will be given here.

This classification has been especially prone to "surprise situations." The first race riot of major size in the United States occurred in Detroit in 1943, and almost 20 years later the burning of a major area in Detroit occurred for roughly the same reasons. The Kent State confrontation between armed forces and students, with its deep and lasting effects on our culture, is a further example. The ability to identify—let alone predict—the occurrence of such events is questionable. It perhaps is an indication of lack of responsive attention by those in authority in the past. Perhaps the country has become more socially conscious, and as a result fewer and less severe "surprise situations" of a social nature will occur in the future. This cannot be taken for granted, however. Alternative situations involving social forces, both logical and illogical, should be considered in evaluating new business opportunities.

Political

Whether or not a force should be classified as political can become highly debatable. If one recognizes that classification is for convenience, most arguments will be eliminated.

Certainly, *nationalization of industries* is a political act. Because of the rapid internationalization of industry, a major concern of any evaluation of a business opportunity in a foreign country has to be the political stability of its government and its commitment to a climate amenable to business. Chile's nationalization of multinational coproration's plants and properties is an example of political forces influencing business opportunities.

Controls legislation in such areas as desalination of water, environmental protection, and food and drug administration impact directly on business. Politically motivated to a large extent, such controls take away various degrees of freedom from management to implement and operate new businesses. The delay in approval of locations of nuclear energy installations has contributed to the developing energy imbalance. The requirements for cleaner air are hindering electric utility expansion and increasing costs in present installations. As a result of these and other forces, solar energy with its potential business opportunities has come to the forefront of serious scientific consideration.

The inclusion of *labor* as a political force is largely because of labor's own active role in politics. Transportation strikes at docks, piers, terminals, of over-the-road truckers, or railroads, and of ships must be considered in evaluating new business opportunities. There is a natural reluctance in purchasing from a single source of supply especially if the producer has a history of labor-initiated work stoppages. This force needs consideration in make-or-buy situations. The movement of the textile industry from New England to the South is an example of labor's effect on business opportunities.

Antitrust legislation and its enforcement is a political force of immense significance. The legislation itself stipulates in general what is and is not legal action. The courts interpret the law and by their interpretations have many times been accused of making law. In the area of new business opportunities, thorough review of this force is required for every case.

The question of "surprise situations" can be illustrated by the following. The Suez Canal has been closed twice in recent memory. This political action was seized upon as an opportunity to construct and operate fleets of supertankers that made the canal itself obsolete. This act made one area of business obsolete and fostered the growth of a new business.

The destruction of a bridge in Zambia was an unpredictable human-caused incident, and yet it influenced the copper business for months.

Technical

New scientific knowledge or known facts applied in a new way put pressure on old businesses and create new business opportunities. The field of fiber optics is rapidly impinging on the communications market for copper wire. The advent of process engineering know-how permitting construction of 1000-ton-per-day ammonia plants spelled the end for many small ammonia plants. The Pilkington float glass process is being adopted by U.S. producers. Electrolytic tin plate replaced hot-dip tin plate.

Birch-bark canoes gave way to wooden-framed, canvas-covered canoes, which in turn gave way to aluminum canoes, then fiberglass canoes, and now ABS plastic canoes. The last three changes have occurred in the past 25 years. Twenty years brings tremendous change in technology, and evaluation requires technological forecasting of the highest caliber.

The "surprise situation" can be found in the realm of technical forces, but in this area the surprise comes in the timing. It is present every time serendipity occurs. For example, science will some day find a cure for cancer. This much can be forecast, but when it will be found cannot be forecast. When it is found it undoubtedly will qualify as a "surprise situation."

The surprise can come with a reverse twist when an entire industry is taken by surprise such as happened when Eastman Kodak introduced their pocket-sized Instamatic line family of photographic products in 1972, all at one time.

Because of the long lead time for major new technical discoveries to mature, the existence of research programs or prototypes often gives advance warning of technological change. To the extent such signs are visible they should be included in all evaluations.

Internal Factors

The third broad classification of factors consists of those that are internal in nature. These forces, largely within the control of a company, can be grouped into two classes and reference points or fixed and moving bench marks established.

Fixed Bench Marks

Certain basic reference points exist in every business. Whether they appear as objectives in formal documents or are unwritten, as often occurs, they do exist, and it is important that they are recognized by the evaluator. The following discussion contains some examples. Each evaluation should contain a thorough analysis showing how the business opportunity compares with fixed bench marks of the business.

Kind of Business

Every management has some obvious areas of interest and some of no interest. For example, a particular manufacturing business may not want to diversify into real estate development or insurance though others may do so or already have. It is important to recognize the available alternatives and to reject opportunities falling outside these alternatives.

Size of Business

Company size to a great extent dictates the size of a business opportunity that can be undertaken. For example, small companies find the risks too great and capital demands too high to undertake very large research projects. In a recent case in point, a small and quite prosperous company found that fixed capital requirements for a new business with moderate risk would approximate the company's total cash flow for a typically good year. Management terminated further consideration of this opportunity based on this one evaluation. Others view the size question differently. The best known case in this regard occurred in 1962 when Albermarle Paper Co., with assets of about $47 million successfully acquired The Ethyl Corporation, with assets of about $240 million. Size, generally, is a bench mark that should be evaluated as soon as possible even if qualitatively only.

Orientation of Business

Many businesses have a basic orientation toward one or more of the four business groups: production, marketing, finance, and research. This orientation may be even narrower, involving a segment of one of the four. If this is a *fixed* bench mark (in many cases it will be at least a *preferred* bench mark) business opportunities should be evaluated on the basis of their compatibility with such and orientation.

The importance of this factor is illustrated by a study of how well 1700 companies performed in terms of providing consistent growth and reasonable rates of return during a 10-year period.[5]

Only 30 companies of the 1700 met the criteria. In these 30 there was a predominance of service and consumer-oriented companies. There was in general a simplicity and directness, both in the concepts of business and in their capitalization. The only conglomerate to make this list was ITT. The report concludes that the key to success in the great majority of cases of the 30 was an unswerving and straightforward pursuit of one essential product or service. Other criteria and other time spans would undoubtedly produce different results, but the influence of business orientation seems obvious.

Growth of Business

New business opportunities are frequently expected to meet or exceed certain predetermined growth rates. These may be stated in specific terms (that is, an average annual growth rate of 7 percent per year) or they may be related to some readily available indicator (that is, gross national product or FRB index of industrial production. The initial degree of accuracy in appraising this factor will tend to fluctuate widely and may well introduce a high degree of uncertainty in the initial evaluation. Subsequent evaluations will permit refinement of this factor. New business opportunities closely allied to existing operations will be easiest to evaluate. Totally new and different markets will be the most difficult to evaluate.

Financial Results of Business

All businesses have financial bench marks for evaluating new business opportunities. Depending on the business these may be in the form of:

Return on investment

Cash flow

Profit as a percent of sales

The financial bench marks will be stated in terms of the method to be used in measuring the opportunity (that is, payout, discounted cash flow, return on investment, and so on. These methods are well described in a book entitled *Techniques in Financial Analysis.*[6]

At the outset of a project, its evaluation against financial bench marks has to be made using broad assumptions. Refinement should occur as rapidly as possible but not to the extent that good projects are prematurely rejected.

The cutoff point is a difficult one to identify, and great care should be exercised in evaluating a new business opportunity against fixed financial bench marks.

Moving Bench Marks

Everything up to this point concerning internal bench marks has dealt with fixed points. A standard has been established against which all new business opportunities can be measured.

At any given time, however, a business has a number of opportunities and relatively fixed resources. The situation then develops that one new business opportunity must be compared with another as well as against the standard set of bench marks. Opportunities compete with each other as well as with existing operations for the resources available. Thus, there is a corresponding set of moving bench marks that the evaluator has to consider.

Some projects will be nearing completion while others will be striving for approval to start. In between will be projects in various stages of their life cycle. Every time a decision is made to add, drop, or accept an opportunity an entirely new situation exists. The bench marks move.

The previous discussion has been rather abstract and in some cases theoretical. In broad terms it presents a procedure to follow when evaluating business opportunities. It is quite obvious that not all factors and bench marks need be evaluated for every new business opportunity. Nor is it necessary to go into equal depth on each evaluation. The following discussion will show how the procedure can operate in practice.

In case this all sounds impractical, consider some of the classic cases where management decisions went awry. In the automobile industry the case of Ford's Edsel and General Motor's Corvair are multimillion-dollar examples of business opportunites perceived, developed, introduced to the market, only to be ultimately withdrawn.[7] There is a very high cost connected with new product failure. DuPont's affair with Corfam (artificial leather), Dow Chemical's with Dowguard (long-life antifreeze), and Monsanto's with Krillium (soil conditioner) are examples from the chemical industry. These all point up the need for ever-improving evaluation techniques.

The first class of forces discussed was long-term trends of Western culture. Some of these trends have developed over hundreds of years and are not about to stop or radically change direction in the near future. In a practical sense it is important that these trends be recognized, understood, and thought about when an evaluation of a business opportunity is made. There will be few cases, however, where long-term trends by themselves will control the selection of one project over another. These trends will subtly influence the future. For example, population growth and urbanization are both on the increase. This suggests that attention be paid to investigating shifting market concentrations and increased needs over the expected 20-year life of a product or service. Once the concept of long-term trends is understood, it will fit nicely into practical evaluations.

The second class of forces discussed were those labeled external factors, those over which an individual business has little control. Practically, these need close consideration. Evaluations should discuss the four subclasses (economic, social, political, and technical) and how each will or will not apply to the business opportunity under consideration.

For example, if the business opportunity involved a solar climate-control system, technical and social factors today would seem to be of primary importance, with political and economic factors of secondary importance. However, all four classes of factors would impact on a solar climate-control system, and the evaluation should involve determining to what degree, when, and how.

The third class of forces (internal–fixed and moving), those normally under control by an individual business, need to be carefully evaluated for every new business opportunity. Questions of importance are these: When should an evaluation be made of a new business opportunity? How detailed should an evaluation be? There are no simple answers, but several rules and generalizations are available.

The first evaluation should be made before any significant commitment or expenditure of resources is made. It should be made at the idea stage. Too often basic or applied research projects are funded and allowed to continue without an evaluation of their worthiness as business opportunities. Something that has high technological appeal may offer little, if any, business appeal. Lest this be thought of as undue restriction on research, one that inhibits innovation at its source and stops the flow of ideas, it need not be. Research staffs should have a certain part of their budget to spend as they choose. Beyond this, however, the research programs should be subject to review and evaluation as are other programs in a business.

The initial evaluation of an idea may be a very brief screening operation against obvious fixed bench marks utilizing existing information or that obtainable with a minimum amount of work. This will probably occur at the idea stage when no one knows if accomplishment is technically feasible. Nevertheless, some evaluation should be made before committing resources. It is absolutely necessary in a selection process that picks a few from many.

The second and all succeeding evaluations should be made at major decision points. A major decision point is one requiring a decision to continue or to commit added resources, whether they be money, personnel, materials, or equipment. Such points occur, for example, when research budgets are approved, when expenditures for plant facilities are requested, and when marketing plans, staffing, and promotional budgets are approved.

As each succeeding decision point is reached, the sophistication of the evaluation should increase. The financial numbers should become more accurate when based on better marketing numbers and better cost information. The business opportunity should become better defined and its future better understood, One thing to be avoided, however, is overanalysis. This results in what has sometimes been called "paralysis by analysis," where resources are so committed to analysis that little commercial development occurs.

At each evaluation these questions should be answered as well as possible taking into consideration all changes since the last evaluation.

1. The basic idea is? (Define and redefine.)
2. Will it work? (Has it been reduced to practice?)
3. Is it desired? (Will it provide the customer satisfaction?)
4. Is it marketable? (Rate alternatives.)
5. How long a life does it have as a business?
6. Will the financial results over the life of the project be a satisfactory reward for the investment?

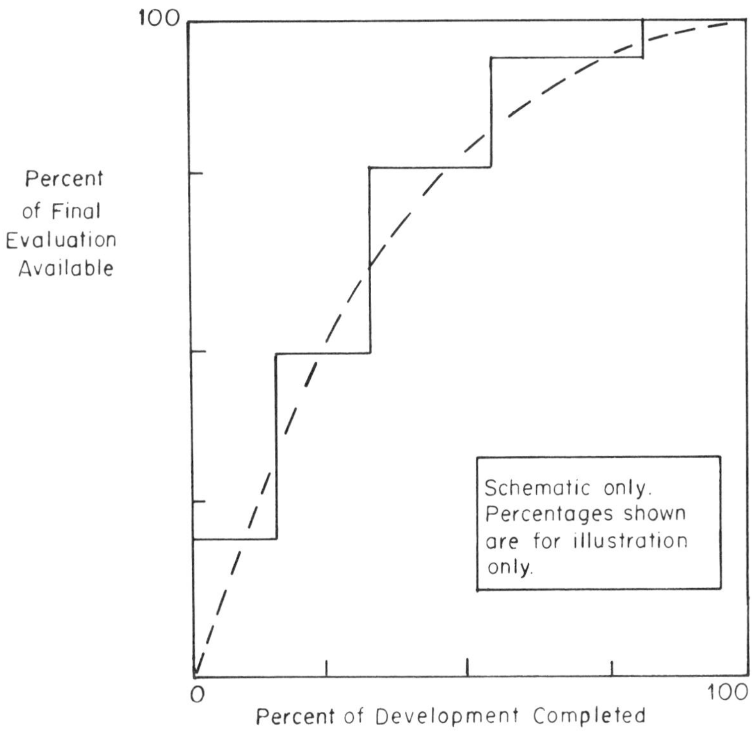

Figure 1. Decision point evaluation procedure.

As a development proceeds, the depth of evaluation will also proceed as illustrated schematically in Figure 1. As indicated evaluations proceed step by step, each one becomes more detailed than its predecessor. It is also interesting to note that the evaluation proceeds ahead of the development. The reason for this may be found in the difference in costs between the two. A small expenditure on evaluation may develop information justifying large development expenses. It can also signal a cutoff point and prevent the expenditure of large sums. Evaluation costs are generally small compared to total development costs.

Finally, there is one last factor, many times overriding all the others, called *business enterprise.* This factor can best be explained by equating it to the old expression "ready, willing, and able." Even when all factors point to a successful development, if management is not ready, willing, and able to proceed—for whatever reason—there is no reason to continue the project.

Summary

All business is concerned with profits, growth, and survival. Because all offerings to society pass through a life cycle of growth and decay, a process of continual renewal is required of business. Presented with more opportunities than their resources can handle, businesses must select those with the highest probability of filling their objectives. Because each opportunity also contains risks, a balance has to be achieved taking advantage of the strengths of a business.

Because every business operates in a continuously changing environment, appraisal of changes and their effects on today's as well as tomorrow's business become important. This is particularly true in selecting business opportunities. For this reason, detailed and sometimes exhaustive evaluation procedures are required.

A three-level evaluation procedure was described involving: (1) long-term trends; (2) external factors normally outside the control of any one business; and (3) internal factors largely under the control of a single business.

Six questions need answering in each evaluation from the initial one at screening and slection time through those made at each decision point. They are:

1. What is the basic idea at this time?
2. Has the idea been reduced to practice?
3. Will the customers obtain some form of satisfaction?
4. Is it marketable?
5. How long a life does it have as a business?
6. Will the financial results over the life of the business provide a satisfactory reward for the investment?

Finally, after each evaluation has been completed, it is necessary to determine once more the readiness, willingness, and ability of management to move ahead and seize the business opportunity as it then exists. Business enterprise has made winners out of average projects, and its absence has made losers out of good projects. A tremendously successful mining and chemical processing operation is being operated today in the western part of the United States by the eleventh company to whom the opportunity was presented. Over the years, the business enterprise of this company made the difference.

Notes

1. "All ideas and offerings which gain acceptance by society pass through some kind of life cycle roughly analogous to the biological cycle of growth and decay." C. R. Wasson. *Product Management* (St. Charles, Ill.: Challenge Books, 1971), p. 24.
2. "In dealing with change you have several alternatives. You can ignore it and let your business die. You can study trends and adjust to them. You can anticipate events and move first, ahead of your competition." E. S. McKay. *The Marketing Mystique* (New York: American Management Association, 1972), p. 157.
3. *A Profile of U.S. Corporate Growth: 1973,* the second in a series of in-depth studies conducted by Midwest Research Institute under sponsorship of Corporate Diversification, Ltd. New York.
4. First propounded by Kahn and Wiener in *The Year 2000,* 1967 and later refined by Kahn and Bruce-Briggs in *Things to Come,* 1972. "The existence of these secular trends is based upon empirical observation rather than theoretical construction and is independent of any particular perspective on—or theory of—macrohistory."
5. Knowlton and Firth, *Shaking the Money Tree,* 1972. Criteria used were: 1960-1970; average annual earnings per share growth of at least 12 percent during this 10-year period and the period 1965-1970; an earnings increase in every year of the decade; an average annual return on total invested capital of 12 percent for both periods.
6. E. A. Helfert, *Techniques in Financial Analysis* (Homewood, Ill.: Richard D. Irwin, 1972).
7. Other cases have been recorded in T. L. Berg, *MisMarketing* (New York: Doubleday, 1970).

Bibliography

Bass, L. W. "Commercial Development as a Measure of Corp. Tech." CCDA Paper 226, Chemists' Club Library, 1964.

Berg, T. L. *Mismarketing.* New York: Doubleday, 1970.

Giragosian, N. H., ed. *Chemical Marketing Research.* New York: Reinhold, 1967.

Helfert, E. A., *Technique in Financial Analysis.* Homewood, Ill.: Richard D. Irwin, 1972.

"How to Evaluate an Investment." CCDA Short Course May 28, 1970, Chemists' Club Library.

"Investment Evaluation." CCDA Meeting 1961, Papers 229-234 Chemists' Club Library.

Kahn, H., and B. Bruce-Briggs, *Things to Come.* New York: Macmillan, 1972.

Kahn, H., and A. Weiner, *The Year 2000.* New York: Macmillan, 1967.

McKay, E. S. *The Marketing Mystique.* New York: American Management Association, 1972.

"The Management of Risk." CCDA Meeting, March, 1965, Papers 287-302, Chemists' Club Library.

A Profile of U.S. Corporate Growth, 1973 Midwest Research Institute and Corporate Diversification Ltd.

Wasson, C. R. *Product Management.* St. Charles, Ill.: Challenge Books, 1971.

8
THE SELECTION AND EVALUATION OF NEW PROJECTS

Newman H. Giragosian
Delphi Marketing Services, Inc.
New York, New York

Carl Pacifico
Management Supplement
Columbia, Maryland

In order to ensure that new product programs match corporate goals, management at most companies establishes a number of guidelines to measure the desirability of various project proposals.* There are a number of advantages to this approach. First, there is a conscious measurement of these proposals against the company's marketing and financial goals. Secondly, those projects that are obviously unsound are eliminated very early so that little effort is expended on them. Finally, these criteria help to instill a sense of logic to the product/project planning process. (It should be remembered, however, that logic is an excellent way to go wrong with confidence.)

These screening criteria are usually described as product, market, and financial criteria. Each will be described in the following sections:

Product Criteria

Product criteria commonly used in evaluating new product proposals are:

Newness
Fit with existing facilities and skills
Proprietary position
Servicing requirements
Technical feasibility
Legal considerations
Organizational support

*The material in this section is by Newman H. Giragosian and was drawn from "Evaluating New Product Proposals," by E. Patrick McGuire, CBR 604, 1973.

Although relatively few products that reach the marketplace are actually new, it is the perception by the consumer that is really important. Product differentiation, in other words, is the element that established "newness" of a product.

Because new product introduction costs can be large, there is a natural preference for products that can be fitted into the company's skills and physical facilities. It is also extremely helpful if the product can be marketed through the company's existing channels of distribution, or channels that are familiar to company personnel.

Market Criteria

Marketing criteria commonly considered are:

- Market size and expected growth trends
- Market positioning
- Effect on existing product line
- Competitive status
- Distribution characteristics

One of the most important factors is naturally the present and forecast size of the market for a product. The importance of gauging the current and future markets for a product cannot be overstated and has traditionally been one of the most important functions undertaken by the marketing research departments. A corollary to determining the size of the market is the problem of determining what share of the market a company may hope to capture with its new entry. Here companies often use an arbitrary assumption, such as hypothecating a certain penetration, say 20 percent, within 5 years. These are dangerous assumptions. A sophisticated marketing research effort can better pinpoint the probable degree of penetration by interrogation of purchasing personnel while carrying out the marketing research effort and by careful analysis of existing customer-supplies relationships.

Another problem associated with market size determination is that of forecasting the probable life cycle of a product or group of products. Because of the trend toward shorter life cycles, the risk of new product introduction is becoming progressively higher. There is the added risk that the product may become obsolete before it even reaches the market. A pharmaceutical firm spent 2 years developing a new drug to relieve one of the major symptoms of an allergic disorder. By the time the drug was developed, tested, cleared by government agencies, and marketed, it was obsolete. A competitor had introduced a multi-effect drug that relieved all of the symptoms of the allergy.

Another problem is that of defining the market for a product, because the market limits for a new product are often anything but clear. For instance, a company might list all the possible applications for a certain product and consider the total market for these applications to be the market at which they should be aiming, whereas other companies might take a much more conservative approach.

Market positioning is an important consideration when the product to be introduced is proprietary in nature and, as is usually the case, the market can be segmented in some fashion. Examples of segmentation are breakdowns by type of users or applications for the product and demographic characteristics, among others.

A new product will also have an effect on the existing product line, and the impact will be either positive or negative. Ideally the new product will flesh out and strengthen the product line, as is often the intent. In other cases the product may be the precursor of a new family of products, and thus it either reinforces the company's market position in a certain business area or helps it move into a new or allied business area. Sometimes however, the new product will have a negative impact; it will take sales away from other products, depending on where they stand in their relative life cycles or what are the price/performance characteristics of the product. If the product (or products) suffering the negative impact is in the last stages of its life cycle, then the negative impact is of little or no moment; but if it is not, then the company stands to lose money on its investment in the former product.

Still another consideration is that of competitive products. Consider, for instance, the situation where the total natural market for a certain series of products with similar end uses is 100 million pounds.

The intrusion of a new product aiming at the same uses must take sales away from the other products because the market cannot expand to accommodate all the products (except by normal expected growth). However, the pressure to meet competitors new product introductions may force companies currently in the marketplace into the position of matching them, thus compounding the competitive situation.

When a product is introduced, a number of factors must be considered:

1. The types of products the new product will compete with in terms of resources and market share attitudes.
2. The nature and strength of the competitor firms.
3. The forms of indirect competition to the new product, and whether they will dilute the market for the product.
4. The lead time advantage of the product before other new products appear.
5. Competitive products' weaknesses or weaknesses in competitors' marketing programs.

6. Whether the possible new competitors are the company's own customers, as is often the case.

Still another important consideration is that of how to distribute the product and how much it costs in terms of money, human resources, and effort. Needless to say, it would be pleasant if the new product's natural distribution channels were those already utilized by the company. Such is not often the case, however. The following questions should be considered:

1. What is the product's normal channel or channels of distribution?
2. Do these channels vary by geographic region or by type of user?
3. What is the user breakdown—that is, what percentage of the total number of customers account for what proportion of sales?
4. Would it be feasible to sell directly to the larger customers, relying on distributors to handle the rest of the sales?

Financial Criteria

Just as important as product and market criteria are the financial yardsticks. These include:

Return on investment (ROI)
Overall profit contribution
Total investment requirement
Profit/risk ratios
Effect on cash flow
Accessory income possibilities

Of the above, return on the investment is the principal means of determining the financial possibility of a project. Much has already been written on the subject of ROI, and it is not possible to repeat this discussion here. The obvious advantage of the ROI method is that it is widely used in capital budgeting decisions and is familiar to many people. Also, as is well known, DuPont has used ROI techniques both for screening new products and in measuring their efficiency. There are two widely used methods for determining ROI: the payback and the present value methods, of which the latter is the more widely used.

Present value methods of ROI include the net present value method, the internal rate of return method, and the profitability index.

In the net present value method, the cost of the investment is subtracted from the present value of the expected future cash flows (which are discounted at the estimated cost of capital). The result of this calculation products an estimated dollar amount that, if positive, indicates that the project should be accepted; if negative, the project should be rejected. Projects with higher positive dollar amounts outrank those with lower present values.

In the internal rate of return method, the present values of expected future cash flows is equated with the investment outlay in such a way that an interest rate is calculated. This interest rate is used to rank the proposals.

In the profitability index method (or benefit/cost ratio), a fraction is calculated in which the numerator is the present value of expected future returns discounted at the estimated cost of capitial, and the denominator is the cost of the proposed projects. If the resulting number is greater than 10 the project should be accepted, and the various projects are ranked according to the amount by which their indexes exceed 1.0.

As an alternative to standard ROI calculations, a "new product rating system" has been developed by Gerlach and Wainwright (see bibliography at end of chapter), which might be applicable where little product differentiation exists and where ROI projections are susceptible to distortion because of short life cycles and by competition.

The formula developed by the authors is:

$$\text{New product ratio} = \frac{d\,(V)\,(Y)\,(P)}{TC}$$

where

d is the coefficient of differentiation

V is the average unit volume

Y is the number of years from commercialization to maturity

P is the average unit gross profit

TC is the total cost of development from product initiation to commercialization

The payback method simply calculates the number of years it will take to recover the investment in a new project; the total new-project investment is divided by the annual average revenue, to give the payback period in years.

Although the most careful scrutiny of the new product and the expected ROI comes at the time of the initial evaluation, there is now an increasing tendency to evaluate the ROI at several different times before the product is finally marketed. These usually come when certain preset sums of money have been expended on a project.

Other problems associated with ROI calculations are the determination of the time period over which the ROI is to be measured and the problems associated with measuring the investment cost. Another factor to be considered is the product life cycle and the corollary problem of making realistic measures of product life and integrating them into the ROI calculation.

Proposal Checklists

Proposal checklists in one form or another are used widely in evaluating new-project proposals. An obvious advantage of a written checklist lies in the fact that it systematizes their periodic review and forces the thinking out of all the factors that need to be considered in their review, as well as eliminating any misconceptions. As an example, the R&D department of a company was under the impression that the company had a policy against licensing for manufacture or sale products originally developed by other companies. However, there was actually no such restriction. Additionally the use of checklists serves to help document the history of the development effort.

Often the checklists are modified so as to create a weighted checklist rating or weighted scoring chart. Specific criteria are given different weights depending on their importance, and the sum of the weighted scores is divided by the sum of the weights to obtain the composite score. Figure 1 presents an example of a checklist.

Concept Testing and Prototype Evaluation

In recent years more industrial product manufacturers have been embracing the idea of concept testing and prototype evaluation, which have been employed for many years by consumer goods manufacturers.

Pretesting has the obvious advantage of providing a checkpoint for deciding whether to continue the program, abandon it, or modify the direction that the development program should take. The type of pretesting selected will, of course, vary with the nature of the product, its market, secrecy requirements, number of potential customers, normal channels of distribution, and other factors.

One of the major problems associated with concept testing lies in the fact that if the testing is done at a very early stage, which would be desirable from the standpoint of providing program direction, it is more difficult for the respondent to react meaningfully to the concept that is being presented. Although it is naturally difficult to accomplish, it is possible to gain some insights as to the probable commercial success of an abstract idea.

Concept testing is very similar to the typical in-depth study performed by the company's marketing research department. In it, they attempt to identify:

1. The range of applications for which the new product might be used
2. The types of users for the proposed product and their specific needs
3. Desired product characteristics
4. The relative importance of the proposed features and the intended functions of the product by the different types of users contemplated for the product

a. Evaluation Form

PROJECT X
PROPOSAL ____ TITLE Connector
PROJECT, PROPOSAL NO.: 1234 EVALUATED BY: NNN ON: March 11
CUSTOMER: ____ MARKET: Commercial PRODUCT: Electronic

1) PROJECTIONS

a) SALES

	1st YEAR 1970	2nd YEAR 1971	3rd YEAR 1972	4th YEAR 19	5th YEAR 19
Sales of new product	$ 200,00	$ 300,000	$100,000	$	$
Lost sales of replaced product	40,000	20,000	–		
Net added sales	$ 160,000	$ 280,000	$ 100,000	$	$
Average sales per year (1)	$ 180,000	$	$	$	$

b) GROSS PROFIT MARGIN

Estimated product cost	$ 3.50 /Ea.	Source	Product Estimate 8775
Predicted selling price	$ 9.00 /Ea.	Source	Sales Forecast by J.D.
Added sales of exist. prod.	$ –	Source	
Gross margin (MG) = $\frac{(9.00-3.50)}{9.00}$ =	.61		

c) DERATING FACTORS

Probability of marketing success	.80	Source	Sales Forecast by J.D.
Prob'y. of Eng. success	.70	Source	Eng'g. Estimate 6789
Life expectancy – years 3	.75	Source	Forecast - Std. Derating
Combined derating factor (Sc)	.42		

d) PROJECT COSTS

	High Estimate	Low Estimate		
Engineering Develop.	$10,000	4,000	Source	Eng'g. Estimate 6789
Production Tooling	$10,000	8,000	Source	Product Estimate 8775
Inventory	$ 3,000	3,000	Source	
	$ –	–	Source	
Total Project Costs (C)	$27,000	15,000		

2) MANAGEMENT CONTROL FACTORS

Sales Volume	.9	Less than preferred volume
Engineering Capability	1.0	
Marketing Capability	1.0	
Facilities Capability	.9	Need Special Molding Press
Market Direction	.8	Special Application
Combined Mgt. Control Factor	.65	

Figure 1. An industrial product manufacturer's proposal evaluation form (courtesy of the Conference Board).

b. Explanatory Notes For Evaluation Form

PROJECTIONS

Sales

Sales of new product. Four years is considered to be the "norm."

Lost sales of replaced product. This is a factor in cases where an existing product is to be replaced by the new product.

Average sales per year. Summation of total net added sales divided by the number of years for which these sales are estimated (1, 2, 3, or 4 years).

Gross Profit Margin

Estimated product cost. A "target" cost is obtained from the marketing organization.

Predicted selling price. This is obtained from the marketing organization.

Added sales of existing products. In some cases, the sale of a new product will cause the additional sales of one or more existing products.

Derating Factors

(1) *Probability of marketing success.* Rating scale in which 1.00 is perfect. Factor is an expression of confidence in the estimate of average sales per year.

(2) *Probability of engineering success.* Rating scale in which 1.00 is perfect. Factor is an expression of confidence in the ability to produce the item at the specified cost or lower, and to meet or exceed technical requirements.

(3) *Life expectancy – years.* Four rating factors as follows:

4 years life expectancy = 1.00
3 years life expectancy = .75
2 years life expectancy = .50
1 year life expectancy = .25

Combined derating factor. (1) x (2) x (3)

MANAGEMENT CONTROL FACTORS

The following factors are each rated on a scale in which 1.00 has been assumed as the norm for purpose of illustration only.

(1) *Sales volume.* Reduced if estimated sales volume of the new product appears to be outside a predetermined, optimum range. This optimum sales volume varies from product to product.

(2) *Engineering capability.* Reduced if the new product should for some reason fall outside the company's current engineering capabilities.

(3) *Marketing capability.* Reduced if the new product could not be marketed through the company's existing marketing organization.

(4) *Facilities capability.* Reduced if the proposed product should require additional capital facilities.

(5) *Market direction.* A rating expressing the relative desirability of the market in which the proposed product would be sold. The company sells to a number of major markets, each of which is rated according to its relative attractiveness. The relative attractiveness of a particular market may change from time to time, and this change is reflected in the assigned rating factor.

Combined management control factor. (1) x (2) x (3) x (4) x (5)

Figure 1. (continued)

5. The design parameters for the product
6. The cost/performance characteristics of the product
7. The marketing characteristics for the product—that is, distribution channels, customer breakdown by size, packaging, geographical location of the customers, and any other marketing information that would help in developing a marketing program

In selecting the people to be called on in the concept test, the company often chooses its own technical and marketing personnel to evaluate the product. More importantly, however, it is necessary for the concept test to be done outside the organization even though, for reasons of secrecy, it may be preferable to keep the entire evaluation in-house.

By having the concept evaluated by the company's distributors and by potential customers, it is obvious that a more objective appraisal of the product concept is made possible.

Although the pretesting of industrial products is performed usually by the marketing research department, it can also be done by the company's sales staff or distributors, as well as by the technical service representatives. Pretesting can be done in a variety of ways, such as personal and telephone interviews and questionnaires, as well as by actual trial use. The method selected will vary with the nature of the product and the market situation.

In order to carry out successful concept tests, the interviewer should make certain that the functions and characteristics of the product are clearly understood by the interviewee and that the various situations in which the product is used as well as how it would be used are described. If a similar competitive product is already in existence, the new product's characteristics should be compared with is, and the products advantages and disadvantages enumerated. Although it may be premature, an expected price or price range should also be given.

Obviously, the personal interview is highly desirable because of its flexibility and the possiblities it offers for an in-depth interview. It is also possible during the free-flowing interview to obtain additional information about the market, competitive products, customer-supplier relationships, customer needs, the parameters of an "ideal" product, and so on.

Although the cost of a personal interview may inhibit carrying out large numbers of them, the personal interview technique can very well be supplemented by the use of telephone interviews and by mail surveys. These two techniques, though not providing the depth of analysis afforded by the personal interview, still allow the gathering of large numbers of responses at relatively low cost that can be used to buttress or confirm the findings gleaned from the personal interviews.

Because on of the major problems in concept testing is the difficulty of conveying an abstract idea to the respondent, the use of demonstration aids is extremely helpful. This will help to prevent the respondent from thinking in terms of existing products or technology.

To summarize, concept testing offers the following advantages and disadvantages:

Advantages

1. It allows early judgment of market success.
2. It makes possible evaluation of the relative merits of several new product proposals.
3. It gives management a checkpoint at which to decide to stop, proceed, or revise the direction of a new project.
4. It can be used to determine at an early stage:
   a. Desirable product charcteristics
   b. Markets and applications
   c. Pricing possibilities
   d. Marketing characteristics
   e. Sales and promotional problems
5. It provides product blueprint or guidelines.

Disadvantages

1. The risk of premature disclosure of company plans.
2. It is expensive and time-consuming.
3. It is difficult to visualize abstract idea.
4. Respondents may overstate actual interest or, conversely, understate it so as to avoid possible commitment.
5. Potential market assessment is suspect.
6. Test results are tentative and inconclusive.

Testing the product usually takes place just prior to the market testing phase. Most of the objectives are technical and relate to obtaining performance and reliability data, and so forth. Some marketing objectives obtained during this phase would be the determination of key purchasing influences, the most effective sales approach, and testing price possibilities. The advantages and disadvantages of product use tests are similar to those for concept testing.

Advantages

1. The tests offer an opportunity to evaluate the product under actual operating conditions.

2. They reveal possible weaknesses.
3. In case of product failure, tests make it easier to withdraw product from the market.
4. They provide useful information on marketing characteristics and technical service requirements.
5. They provide possible basis for validating performance and economic claims sponsor may make for the product.
6. They assist in future promotion of product by providing possible source of testimonials.

Disadvantages

1. The tests involve risk of premature disclosure.
2. They tests are costly and time-consuming.
3. It may be difficult to find suitable variety of tests.
4. Test situations may differ from actual market conditions.
5. The customer may misuse the product.
6. Only a limited quantity of the product can be tested, and product tests are not always susceptible to close observation and evaluation.
7. Users' reports may be biased without the sponsor being aware of it.

Use of Mathematical and Statistical Models for New Product Screening

Models, although in use for macro business and government planning activities for at least several decades, have been used only sparingly for new-project evaluation and development. The use of models was discussed in the book *Chemical Marketing Research* (refer to bibliography at end of this chapter) in which it was pointed out that relatively few marketing research projects are susceptible to analysis by model building. The major problem with the use of models is that it is difficult for a static abstract representation of the dynamics of the marketplace to be current and realistic. To be useful, a model should be automated and interactive, and the cost of setting up the model and obtaining the necessary inputs becomes prohibitive, or at least makes the building and maintenance of a cost-effective model extremely difficult. Generally speaking, a model is not cost-effective if the length of time needed to build the model equals the useful life of the model. This section describes and summarizes the current state of the art in the use of models.

There are two general types of models: descriptive models and decision models. The former attempt to describe a particular phenomenon such as product distribution or behavioral change and are concerned with simulating and testing relationships between a particular variable and others within the system.

Decision models, which are the ones that are used for new-product or new-project planning, are designed to determine the probable results obtainable from different values. In the construction of such a model various assumptions are made about how the system performs.

Most of the models used today in product-screening work are nonlinear in nature; that is, a change in one of the variables does not have a constant proportionate impact within the model.

Marketing models may be classified further into stochastic and deterministic types. Stochastic models take into account the varying probabilities of outcomes, whereas deterministic models assume the certainty of particular outcomes or events.

A relatively popular type of model used in new product screening efforts is the risk-analysis model. This type, in essence, provides a basis for measuring the degree of uncertainty surrounding each type of new product question such as demand expected, probable share of market, investment return on investment, and product life. In risk analysis the laws of statistical probability are applied to the key variables affecting the expected return on investment. Using risk-analysis techniques, it is possible to construct a range of estimated values for each of the major factors that influence the new product's chances of success, as shown in Figure 2. Then, by using random probability techniques, a single value is selected within the range for each of the factors. All of these random selections are then combined in an appropriate ROI equation, and the probable rate of return can then be calculated. By repeating the selection procedure of ROI computation, it is possible to obtain a listing of various ROI rates for a proposed new product and to determine the probability of these occurring. A specific return can usually be achieved through more than one combination of events.

By supplying several different estimates of variables, such as a high estimate, a low estimate, and the most probable one, it is possible to measure the ROI probability distribution at various levels of each of the variables. These can then be combined in a graphic form so that one obtains a combined ROI distribution probability. The configuration of the curve also supplies an idea of the overall confidence in the success of the new product. If the curve is flat, the forecast reveals a high degree of uncertainty; a bell-shaped curve indicates a high degree of confidence in a particular rate of return, as shown in Figure 3.

The type of analysis just described naturally requires the processing of many computations, so that a comprehensive computer program is required. It should be able to solve not only the individual ROI equations but also the probabilities associated with each of the screening grades.

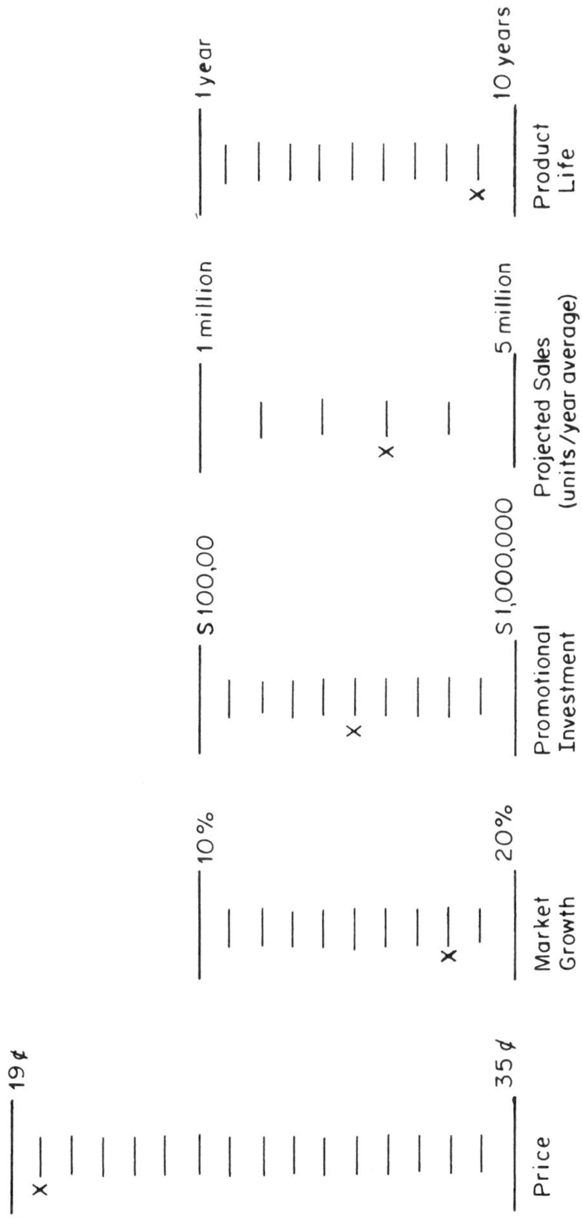

Figure 2. Risk analysis chart.

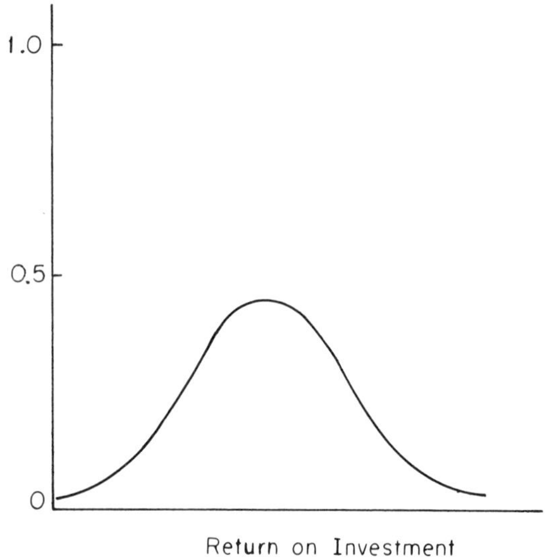

Figure 3.

Some Specific Screening Models

NEWS (New Product Early Warning System) is a screening model developed at BBDO. It attempts to predict a new product's probable performance in a test market. It combines many factors to determine how many people will buy a particular product and is based on the assumption that advertising and promotional effort will induce customers to visit a store, so that the company will obtain a stated level of trial.

SPRINTER (Specifications of Profits with Interaction under Trial and Error Response) is a model developed by Glenn L. Urban of MIT. It is designed for use with an information network, and the model provides decision information on each of the steps in the network. The major elements in the model are demand, cost, and profit. The mathematical relationship among these factors is developed and then put in equation form, which is solved by computer simulation that maximizes the profits of a new product under varying conditions.

There are, however, a number of limitations on the use of screening models, and some companies that have experimented with their use have abandoned them. These limitations are:

1. Too many screening models are oversimplified. The real world is much more complex and subtle than the world that the models are capable of depicting.

2. The marketing models are not as objective as they seem, because they are subject to the individual biases of the designers.
3. Developing these models is costly and time-consuming; this expense often cannot be justified, especially when a company has relatively few project proposals to evaluate.
4. Finally, most screening models are not applicable to early-state screening.

Project Selection*

Introduction

The single most important factor in the success or failure of any project is its selection. Appropriate projects move ahead steadily, encountering only minor reverses, whereas inappropriate projects seem plagued with unexpected difficulties, resulting in delays, increased costs and reduced returns, if not total failure.

A project should not be examined in a vacuum but in its overall relationship with the company that plans to undertake it. This implies that a project is not "good" or "bad" in itself; it is simply appropriate or not appropriate for a specific company at a given time.

The project must be evaluated against all other projects available to the company at that time, so that even a potentially profitable project may be rejected in favor of a more attractive one. Indeed, if not suitable project exists, there is always the alternative of doing nothing, husbanding the company's resources while searching for an appropriate project, rather than accepting a marginal one simply because it is available.

New Activity Objectives

Integrating the selection of the new project with current company activities is straightforward enough, but it does require some analysis and planning.

It is assumed that the company has written objectives that are quantitative, as to performance desired and the time this is to be achieved. Through forecasting, the contribution to be made by current activities to these objectives can be estimated. The difference between the overall objectives and that portion probably available through current activities constitutes the minimum *new activities objectives* (NAO).

These NAO cover more than the basic financial objectives. That is a potential project cannot be viewed simply as a banker's investment; it must be part of an overall corporate development program, which involves some

*This section was written by Carl Pacifico.

selected balance of long- and short-term projects, risky and secure projects, major and minor projects, and possibly other aspects of healthy growth and diversification. For example, NAO might include the following:

1. Within 3 years, achieve sales of not less than $100,000 each in at least two markets not currently served.
2. Be based on technology other than current major activities (for example, chlorination, esterification).
3. Principal raw materials must not include natural fatty acids (to diversify away from a raw material that consistutes too large a portion of our present total purchases).

Through setting up such NAO, the company not only interrelates the new project with its present activities, but sets the direction in which it wishes to move the overall corporation.

Project Size

Limits on the size of appropriate projects may or may not be included in the NAO, but the subject should be considered carefully. There is seldom a problem on the maximum size of a project, because it is readily recognized that there is some size the company simply cannot handle. An increasing number of companies are setting limits on the minimum size of projects they are willing to consider. This limit is often the smallest size that would have a significant effect on corporate profit, usually about 5 percent of total sales for a new field of activity. This is the very least that should be done in selecting project size, because it is rather ridiculous to see major companies utilizing scarce management talent to work on projects whose effect would be lost in rounding off the totals in their balance sheet.

However, this leaves open the question as to whether there is some optimum size. A theoretical model has been set up, assuming that a company makes a series of investments, each one a relatively constant percentage of its net worth at the time, but further assuming that it would not invest all its assets in one project so that failure would wipe out the company. This theoretical model results in an optimum for project size of the following type:

$$x = \frac{pR - qL}{RL}$$

where

x is the project size as a percentage of available net worth

p is the probability of success

q is the probability of failure

R is the profit on successful investments, as a multiple of x

L is the loss on unsuccessful investments, as a multiple of x

The expected value of capital (C) compared with initial capital (C_0) for a series of n investments of this size would be:

$$E[C/Co] = 1 + \frac{(pR - qL)^n}{LR}$$

Calculation will confirm that no other size of investment would yield as high a capital under these conditions.

Although the model is obviously simplified compared with most business situations, it does indicate that most companies choose projects too small in relation to their assets.

Number of Projects

Most companies of any size work on many projects simultaneously. However, although it requires no particular skill to start a project, it is much more difficult to complete one successfully. Because so many projects tend to drag, there is a question of whether there is an optimum number of projects, beyond which efficiency is reduced.

This does appear to be the case, and although the optimum number cannot be calculated quantitatively, there is a roundabout way of estimating it. The relationship between effort and results on any project will have the general form shown in Figure 4.

That is, considerable initial effort achieves relatively small results. This is followed by a period of rapid increase of results in terms of effort. Beyond that, it takes a great deal of effort for each incremental gain in results.

Some quantitity of results constitutes success of the project. This, in turn, is associated with some amount of effort. Any less effort results in failure of the project.

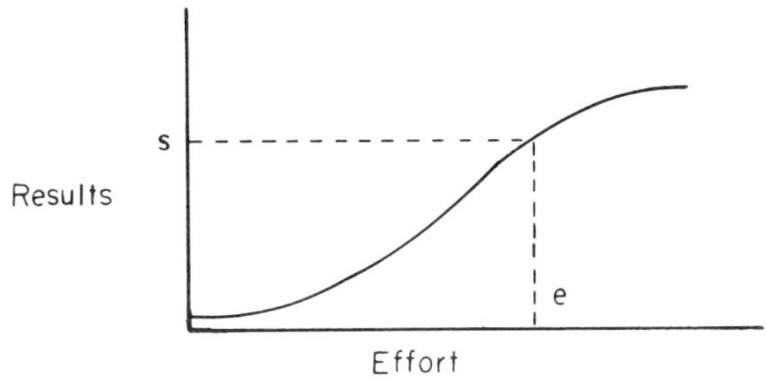

Figure 4.

It is theoretically possible for a company to distribute its available strength over so many projects that no one of them receives enough effort to achieve success. This situation is unfortunately fairly common; there is always "progress," but never "success." A better procedure is that of ranking available projects into some order of priority, either intuitively or through the techniques to be described. Although it is not possible to know exactly how much effort is "enough" for each, this can be estimated, allowing some margin for safety. The strength required for each project is then assigned successively in order of desirability of the project. When the strength available to the company runs out, all less desirable projects must be set aside, temporarily or permanently.

This is a rather long preamble to the point, but it is important. If there is not enough strength available, whether existing or readily assembled, to handle a project, it should not be undertaken, regardless of how desirable it may be.

Basic Relationship

All projects, whether a new plant for synthetic fibers or replacement of a pump, involve the same basic relationship.

$$\text{Risk} \times \frac{\text{Total benefits}}{\text{Total investment}}$$

This is the ratio of the total benefits to be obtained, multiplied by the probability of achieving them, divided by the investment required to get them. Expressed otherwise, it is a measure of the total rate of return on the project modified by the probability of getting it. When expressed quantitatively, this is an absolute index of the desirability of a project.

Every project analyst handles this relationship intuitively if not quantitatively, so there is no need to labor the point. However, it does show clearly that projects are of four basic types:

1. Low risk, low return
2. Low risk, high return
3. High risk, high return
4. High risk, low return

Because low-risk, high-return projects are rather rare, most companies concentrate on low-risk, low-return projects with some portion of their assets devoted to high-risk, high-return projects to improve their average. Nevertheless, an examination of research projects will show a suprising number of high-risk projects whose return, if the project was successful, would hardly cover the investment required.

This basic relationship also shows that the three factors that comprise it are equally important in establishing the desirability of the project. There is little point in extremely precise calculation of one factor, such as investment, while analyzing risk very roughly, if at all.

Expanding the Basic Relationship

For convenience of analysis, the basic relationship can be expanded as follows:

$$\text{Risk} \times \frac{\text{Annual volume} \times \left(\begin{matrix}\text{Unit}\\\text{Price}\end{matrix} - \begin{matrix}\text{Unit}\\\text{Cost}\end{matrix}\right) \times \text{Life} + \text{Salvage value}}{\text{All investments}}$$

There are several important advantages of an analysis of this type:

1. There are no "research projects" or "marketing projects," but only overall corporate projects. The overall project must be sound before any substantial money is spent on any aspect of it.
2. It obliges the company to gather data on all aspects of the project before undertaking it. Where such data are not available, such as probability of technical success or probable sales volume, it provides an indication of how much is worth spending to get them.
3. It is objective. Disputes as to the desirability of a project can be shifted from overall subjective impressions to disagreement as to the validity of the numbers assigned the factors, which in turn can be investigated objectively.
4. It provides a means of policng the progress of an accepted project through review of any changes in its components.

Although the investments are not the same for all projects, each of the following items must be examined and assigned a probable cost, if applicable:

Marketing research

Laboratory research

Market development

Engineeirng

Plant cost

Start-up losses

Working capital

Legal

The total investment in a project is thus always more than the direct plant costs and, in the case of specialty products, may be many times higher than plant costs.

Some projects are considered to achieve other benefits for the company in addition to those tangible benefits listed above. As an example, a company may consider it worth $100,000 to obtain a market in a given industry, based on future benefits that should develop. This is easily handled by adding this sum to the numerator to account for intangible benefits.

Before examining the individual components in the relationship, it is important to note that there are a substantial number of interrelationships among the factors. These relationships should be explored in each project to select the conditions at which the project would be most attractive. The most important of these relationships are:

1. Price vs. volume
2. Volume vs. unit cost
3. Volume vs. investment required
4. Research investment vs. plant investment (additional research may result in lower plant costs)
5. Research investment vs. unit costs
6. Technical risk vs. research costs (additional research may reduce the possibility of failure)
7. Commercial risk vs. market development cost
8. Technical risk vs. plant investment (increased investment may provide some insurance against failure)

If the resulting ratio is not attractive under the most favorable conditions foreseen, the project should obviously be abandoned. As the project proceeds, the ratio should be reviewed periodically, with the data becoming more reliable in each review. If ever the ratio become unfavorable, as through a change in market conditions or an unexpected technical difficulty, the feasibility of continuing the project should be studied carefully. If a satisfactory ratio cannot be restored, it is time to gather the courage to quit.

Comments on Factors

The data required to evaluate the project are obtained from internal estimates and standard marketing research procedures. However, a few comments may be helpful.

Volume

The total market, particularly in the case of commodity products, can usually be determined with reasonable accuracy. The problem is determining what percentage of it a specific company can get. If there are several similar, but distinct, products competing for the same market, as in the case of films for packaging, the techniques of Pender can be applied. More recently Kesting has proposed a formulation of the share of the market that can be expected to be gained by a new supplier.

This still leaves the problem of how the various companies serving the same market with the same product will share it. One approach is that of taking successive approximations of what can be achieved, as follows:

100 percent probability of 1,000,000 pounds	=	1,000,000
80 percent probability of another 2,000,000 pounds	=	1,600,000
50 percent probability of another 2,000,000 pounds	=	1,000,000
20 percent probability of remaining 3,000,000 pounds	=	600,000
Probable volume	=	4,200,000

Unit Cost

It is now recognized that unit costs for commodity products decline steadily with manufacturing experience. This should be taken into consideration in making the analysis.

Selling Prices

Selling prices of commodity products tend to decline at the same rate as unit costs. Very often, operating profit, as a percentage of selling price, remains relatively constant. It must be noted, however, that this is a constant percentage of a declining value, and because overhead costs are constant at best, the overall result is a narrowing profit margin with time.

It is often possible to calculate the rate of decline in price because this rate is sometimes surprisingly constant over a long period of time. For example, the selling price of sodium lauryl sulfate fell at a rate of 4 percent per year for more than 15 years; at one point there was a price increase, but the next tumble brought the price right back to the trend line.

The prices of specialty products, however, are related more to value contributed to the use than to costs, so that price projections must be made on that basis.

Project Life

This item is seldom given adequate attention, probably because it is so difficult to quantify. Nevertheless, even a fair estimate is better than ignoring it. A project earning even 30 percent a year is a poor one if it last only 2 years.

Care must be taken to avoid the trap of assuming an indefinite life for some commodity projects. The product itself may continue indefinitely, but there may be changes that make the *specific project* obsolete. For example, there may be changes in quality requirements, a less expensive new process, or a dramatic change in costs through much larger plant sizes.

The manufacturer of specialty products is faced with the same problem, but at least he knows his project will be obsolete sooner or later and so makes some estimate of a reasonable life that would provide an acceptable return. Estimates can be based on general experience in a given field (for example, metal-cleaning formulations have an average life of 5 years) or may be tied to changes in the market for his product (for example, if a lubricant is designed for a given aircraft, he can estimate how long it will be before that plane is obsolete.

In general, most products manage to survive a little longer than expected.

Salvage Value

There is little problem in calculating this factor; it usually consists of recovery of working capital and land plus whatever can be used or sold of the plant equipment. However, this can influence the investment decision as to what type of equipment should be installed. More flexible equipment will cost a little more but can be converted to other products when this one dies.

There may also be some salvage value in a market position and laboratory equipment, but it is probably better not to include this in the calculation.

Investment

One problem to be noted is that the benefits accrue over some period of time in the future, whereas the investment must be made at some earlier time. It is therefore necessary to bring these factors to the same basis. This can be done by discounting future benefits back to the date of the investment. Alternately, the value of money can be added as interest to the investment over the period of life of the project. The error introduced through omitting this is not usually significant (in comparison with errors inherent in other estimates) in short-term projects but can be significant in longer-term, capital-intensive projects.

Risk

This is the factor that separates the true managers from the number jugglers. Because there is yet no rigorous quantitative measurement of the risks in a project, a qualitative value based on experience and judgment must be used. This lack of precision, however, makes the evaluation of risk more, rather than less, important.

All the risks involved in the project must be identified, because it is usually the unexpected problem that wrecks a project. Where possible, the risks should be eliminated. Some of those that cannot be eliminated should be insured against, such as providing an extra manager as backup for a key person whose loss would hurt the project. And those risks that can neither be eliminated nor insured aginast must be watched warily throughout the life of the project.

Risk Capacity

Quite independent of any specific project, every company has some capacity for handling risk at any given time. This might be defined as the maximum loss it can absorb without encountering serious trouble. Depending on this situation, this might be anything just short of bankruptcy to a loss of 10 cents per share in earnings. Whatever criterion is used, it should be known. Any project involving the possibility of a larger loss simply cannot be considered.

In addition to financial loss, any project failure involves some loss of prestige that may carry over into other activities. The capacity for this type of loss must also be considered. Thus a company may be able to tolerate a loss of a given size in a new field of activity but may be unable to absorb a loss of exactly the same size in its principal field of activity without suffering additional losses on its current products.

Extent of Risk

The overall risk of a project is related both to the number and to the seriousness of the individual risks in the project. Each "if" in a project indicates an area of uncertainty. Where there are no "ifs," there is an order. The more "ifs," the more obstacles to be overcome. And they must *all* be overcome; failure on one is failure on the entire project. Where there are five major risks, it does not matter how well you succeed on four of them, nor does it matter whether you fail on the first or the fifith. It's all or nothing. Also, because each area of uncertainty is itself capable of wrecking the project, the combination of all uncertainties is obtained by multiplying the uncertainties, not adding to them. Thus if 100 percent is scored for each success and zero for each failure, the result of four successful and one unsuccessful item is zero, not 80 percent.

As a convenience in handling this relationship, it is easier to use the reciprocal of the risk; that is, the probability of success, which can be expressed as a decimal fraction, such as 0.9, and so on. The overall risk, can then be respresented as

$$R = R_1 \times R_2 \times R_3 \ldots R_n$$

Not all "ifs" represent an equal degree of uncertainty. In some cases, the probability of success is very high, in others it is very low. Assigning values will be discussed later, but it is interesting and important to note the dramatic importance of the number and degree of risks. This is shown in Table 1.

TABLE 1
Estimating Risk

Number of Risks	Probability of success when R = 0.5	R = 0.1	Comment on R = 0.1
0	1.0	1.0	Success
1	0.5	0.1	Maybe
2	0.25	0.01	Need luck
3	0.12	0.001	Need a miracle
4	0.06	0.0001	Need a replacement
5	0.03	0.00001	Need a psychiatrist

Types of Risk

It is convenient to group the risks according to the principal aspects of the project. For a typical project, these might be:

Commercial risk
Technical risk
Financial risk
Legal risk
Other risks

Each of these categories can then be studied thoroughly. Examples of questions concerning legal risk that might be asked are:

1. How likely is it that an adverse patent will issue affecting the operation?
2. What is the probability of a legal ordinance that interferes with proposed waste disposal method?
3. What is the probability of legal suits by users of the product?

After all the risks in a category have been identified and evaluated, they can be gathered into a measure of the overall risk of that aspect.

Assigning Quantitative Values to Risk

Assignments of quantitative values for probability must still be left to judgment.

In general, technical risks tend to be overestimated and commercial risks underestimated. There are examples of spectacular technical failures, in which plants never operate. More often, however, the technical problems are solved, although it takes longer and costs more than anticipated.

This is not the case with commercial problems. No amount of sales effort, including advertising, can make the product salable. And the passage of time, instead of helping as in the case of technical problems, only compounds the commercial dilemma as others solidify their hold on the market.

Thus if some risk must be accepted, it is usually better to select among equivalent projects those whose risk is principally technical rather than those whose risk is principally commercial.

Defendability

The ability to defend a project is an aspect of "risk," but it is sufficiently important to merit separate treatment.

There are few activities more frustrating and less profitable than opening up a new market only to have it taken away by someone else. No one ever owns a market; there is neither law nor custom that permits you to plant a flag on a new activity you have discovered or developed and so claim it to yourself in perpetuity. It is yours only as long as you can defend it against all comers, whether presently existing or to be organized specifically to try to take it away from you. In short, there is no point in starting a project that you cannot defend if it is successful.

The overall effect of the attack is always a reduction of profit, but it can occur in different aspects of the project and in varying degress of intensity. The vulnerability of each aspect and its method of defense must be examined separately. The attack may be broad and general, as when the farming cooperatives decide to make their own fertilizers. If the company that started the project of offering that particular fertilizer can hold its sales volume at all, it will have to sell the product at a considerably lower price. In other projects, the attack may be selected, as the offering of a higher quality product to a specific market or the selection of a location that cuts off one of your regional markets through faster delivery. At times neither sales volume nor price is reduced, but profit is lost through the unexpected costs of advertising or special services need to hold the market against attack.

The defense required will depend on the specific project, but at the very least an analysis should be made to determine how vulnerable the project is and how likely it is that the defense will hold. Some common types of defense are:

1. *Timing.* In some cases getting there first is itself a defense. This is particularly true of specialty products; once adopted in a use, the purchaser is reluctant to change for a minor benefit because of the risk of failure involved.
2. *Size.* Although not an absolute defense, size does reduce the number of attackers. A large size is beyond the reach of smaller companies, whereas a small size is beneath the interest of major companies. Projects requiring

intermediate amounts of capital and skill are the most vulnerable because all but the smallest companies are potential competitors.

3. *Patents.* A solid patent position is an absolute defense in some aspects of the project, but too often it provides a false sense of security as the project is attacked on some other basis.
4. *Trade secrets.* In complex projects, trade secrets are often more valuable than patents. The problem here is to keep them secret, to guard in particular against employees who leave and set up competing operations.
5. *Raw material position.* This is extremely important in commodity products.
6. *Contracts.* Contracts are very useful but not often easy to get before the project is started.
7. *Commercial position.* An established commercial position in the market where the new product is to be sold can be a surprisingly strong defense. In some fields, newcomers may struggle for years for acceptance of a superior product or an equivalent product at a lower price.

Of course, most defensive patterns will be made up of a number of these individual types of defense. In some cases, it may be less costly to abandon a portion of a market rather than attempt to defend all of it.

Project Champion

No project should be undertaken unless there is a project champion at a sufficiently high level in a line function to support it. No project ever proceeds without encountering some obstacles. Although everyone may claim to have supported a project that eventually succeeds, the sad fact is that during the course of the project, many, if not most, of the participants may have constantly expressed their reservations, as though preparing their excuses so that they will be freed from responsibility should the project fail. Their mutterings reach a crescendo as each obstacle develops: "It was obviously a foolish project to begin with, and we should abandon it before we throw good money after bad." This is the time when a champion is needed who is willing to stake his or her reputation on a successful outcome and has the weight to see that the obstacles are overcome, one way or another. The importance of having a project champion can hardly be overestimated; without one, it will usually be better to defer or drop the project regardless of its attractiveness.

Alternative Approaches

The most widely used method of project selection probably continues to be that of the whim of an executive. This is based on analysis such as "we ought to be in that business," or "If it's a good project for them, it should be a good business for us." Research projects are started on the basis that the technology is "interesting."

Nevertheless, most companies do have some formal basis for screening most, though not all, of their projects. This may be nothing more than allocating $200,000 for research to support a current area of business, with no analysis of how much good can be expected in return. In most cases, however, an analysis is made on projected rate of return on investment, with the tacit assumption that the project will be successful.

The more sophisticated methods of project analysis do attempt to take both return on investment and the risk factors into account, though the latter is handled only indirectly. This is done through use of a checklist of those factors the company considers important in its investments. Such lists may include items like: (1) return on investment; (2) uses present raw materials; and (3) can be sold by present sales force.

The projects are then rated on some scale such as excellent, good, average, below average, and poor. The distribution of the markings on the chart present a rough visual picture of whether the overall project appears to be above average, average, or below average.

Some companies have recognized that this approach is essentially qualitative and would like a more quantitative measure, especially if a comparison is to be made among competing projects. This is then handled by assigning some arbitrary weight for each factor on the checklist and scoring 100 percent for excellent, 50 percent for average, and so on. An overall score is then compiled for each project.

It is my opinion, however, that although the checklist approach is helpful on a qualitative basis, it introduces boobytraps when it is made quantitative. There are two reasons for this. The first is that the only valid basis for weighting the factors is their effect on return on investment, and this is seldom, if ever, done. The second reason is that as shown in the method described, the factors must be multiplied, not added. For example, a project might score "excellent" in every aspect except rate of return, which is zero. The additive procedure would give a score that is almost perfect, whereas the project is obviously worthless.

It would therefore seem desirable to prepare such checklists as a means of estimating the values to be used in the procedure proposed rather than as an independent scoring system.

Summary

In summary, a suitable project must meet *all* of the following criteria:

1. It must fit the corporate objectives governing growth and diversification.

2. It must be of an appropriate size.
3. It must exceed the company's minimum standards for probable return on investment.
4. If it fails, the loss must not be more than the company can tolerate at that time.
5. If it succeeds, it must be capable of being defended.
6. There must be a company champion willing to risk his or her reputation on its success.

Bibliography

Doud, W. C. "Evaluating the Effectiveness of Commercial Development." Talk given at symposium entitled "Evaluating the Effectiveness of Commercial Development," June 1971.

Kelly, Ralph E. "Evaluation of New Ventures Today." Talk given at symposium entitled "Evaluating the Effectiveness of Commercial Development, June 1971.

Kesting, R. Ama Course 5595, 1976.

McGuire, E. Patrick. *Evaluating New Product Proposals*, Conference Board Report No. 604, 1973.

Pacifico, C. *Getting Your Money's Worth in Prosperity*. ACS, August 1960.

——. *Is it Worth the Risk?* ACS, March 1964.

——. *Project Selection*. Rosevelt V. Chemical Economics lectures, November 24, 1969.

——. *Selecting Optimum Investment Size*. ACS, September 1969.

Worth, James H. "Timing in Commercial Development." Talk given at symposium entitled "Evaluating the Effectiveness of Commercial Development," June 1971.

9
NEW PRODUCT PRICING

Newman H. Giragosian
Delphi Marketing Services, Inc.
New York, New York

The price at which a new product is offered for sale is a very important factor in determining whether a new product is a success or failure. It can be so high that important fields of utility are automatically excluded. It can be entirely too low and cause serious embarassment to both the producer and potential customers if drastic increases must be made once commercial production is attained. In the event of a drastic increase, a major portion of the expected market may instantly disappear. Conversely, a substantial reduction may result in large, new unexpected uses and quickly make the new plant capacity entirely inadequate.

New product pricing is largely a matter of judgment. Not only is it one of the most important of marketing problems, but it is usually surrounded by a great deal of mystique which adds to the difficulty of arriving at proper rationales for setting prices on new products. The high proportion of new products that fail in the market place has been attributed to the difficulty of pricing them correctly, and undue emphasis on initial profit relative to realistic cost/performance analysis. Pricing is even more difficult today than formerly because new products today deteriorate more rapidly, competitively speaking, than did products introduced twenty-five years ago.

The purpose of this chapter is to examine the factors influencing the prices for new products and to suggest techniques for setting these prices. The importance of setting appropriate pricing objectives will be discussed, and various pricing strategies will be analyzed.

Factors Affecting Pricing Decisions

First, let us examine the factors that affect pricing decisions, keeping in mind that price should try to satisfy three different needs:

1. The needs of the marketplace.
2. Competition—that is, what will be the reaction of the competition to the price set for the company's new product.
3. The Company's desired profit margin.

The principal factors in pricing new products are:

1. Competitive considerations
2. Market or sales expectations
3. Marketing factors
4. Product characteristics
5. Profit and cost considerations
6. Other factors

It should be emphasized at this point that the first three factors listed immediately above must be satisfied before profit and cost considerations; otherwise it will be necessary to drop the product.

Each of these factors include a number of attributes and other considerations, the weighted sum of which suggest the range of price in which the new product should be placed in order to best enhance its commercial reality and to ultimately provide the optimum return to the company. The many questions and/or considerations under each of the above major headings are listed below.

Competitive Considerations

Anticipated reactions of competitors to changes in price

Availability and prices of competitive products

Availability and prices of substitute products

Cost/performance characteristics of the product

Basic prime behavior characteristics of competitors

Degree to which competitors' pricing strategies have affected your company's sales volume

Market or Sales Expectations

Share-of-market goal or potential

Estimated sales volume for company's new product (near or long term)

Specific market segment(s) targets based on superior cost/performance characteristics

Company's past and current share of market

Sales volume at alternative prices

Marketing Factors

- Degree to which product fills a real or perceived need or requirement
- Number of possible customers
- Degree to which customers emphasize
  - Uniqueness of product
  - Quality
  - Price
  - Service
- Channels of distribution used
- Selling methods used

Product Characteristics

- Value to customers
- Cost/performance characteristics
- Does company offer full or partial product line
- Importance to product line
- Product's expected degree of proprietary strength
- Stage of the product life cycle
- Product life
- Elasticity of demand for product
- Patent control
- Product uniqueness

Profit and Cost Considerations

- Profit
  - Return on investment (or sales)
  - Cost plus profit
  - Current profit levels
  - Relative profitability of each product
- Volume-cost-price relationships for each product in company's line
- Contribution of each product to overhead
- Costs
  - Production costs
  - Development costs
  - Selling or promotional costs

- Administration costs
- Operating costs
- Investment cost to achieve economy of scale

Financial risk

Other Factors

Consistency with product-line price structure

Raw material

- Availability
- Price trends
- Sources

Anticipated reaction of prospective customers

Maximum price obtainable

Stability of industry price structure

Psychological pricing factors

Unused production capacity

Geographical market where product is to be sold

- Domestic (national or regional)
- Foreign

Corporate image implications

Government regulatory impact

Satisfying all the above needs is not always possible. The price finally set that will yield the largest sales revenue reflects the uniqueness of the product and *price/performance characteristics* relative to competitive products.

Needless to say, the price for a new product depends on the newness of the product. Naturally, a product that is unique and new to the world, rather than being new to the company, will probably be able to command a higher price than one that is more of a "me-too" product *and which has competitive products in the marketplace against which the buyer can measure its potential worth* in specific applications.

As stated earlier in this book, there are three classes of new products:

1. The not-thought-of-idea, or the thought-of-but-not-made product that is new to everybody
2. The product that is not new to the world, but is new to your firm
3. The product that is not new to your firm but is new to most of the customers who should buy it.

The price that can be commanded, and the pricing strategy selected, varies for each of these cases.

Defining Pricing Objectives

Before establishing prices for new products, it is necessary to define objectives, a self-evident but often overlooked step. Examples of company objectives would be:

1. To enter new business areas
2. To complete or add to an existing product line
3. To serve as a counter to cyclical or seasonal products
4. To provide a more stable base for the company
5. To maintain or increase the overall profit growth of the company
6. To utilize facilities or materials otherwise unemployed

These goals may or may not be consistent with the desirable goal of maximizing sales revenue. Other factors influencing the pricing decision would be such items as the company's product and pricing strategies. Is the company a price leader or a price follower? To what extent does the company's public image and public relations activities influence pricing strategy? Needless to say, government regulations and laws will also necessarily dictate the parameters of the pricing decision.

It can be seen, therefore, that the pricing decision is arrived at through a series of steps. Initially, production costs are obtained, together with very approximate data on sales volumes attainable at various price levels; these should be obtained from the marketing research group as part of its initial screening study. These are refined at later stages as more in-depth studies are conducted by marketing research.

As will be seen later, the *cost/performance characteristics of the new product are the major factor in helping set the price.* The relative advantages and disadvantages of the product in competition with substitutes *for each application* must be determined, usually *during the market testing phase.* The following questions should be answered regardless of price:

1. How much can be used in application, and for how much volume is it technically preferred?
2. Very importantly, what is the threshold price for each application, that is, at what price would the buyer consider other substitutes? This is a very important point. Too often, research on this aspect is glossed over. Needless to say, however, the proper definition of the threshold price will save (or cost) the company a great deal of money.
3. For what applications would the buyer use the product exclusively, regardless of price?

Answers to these questions are extremely difficult to obtain. Also, in spite of rigorous field interviews with potential buyers and market testing of the product at the customer's facilities, the answers obtained may very well be erroneous or misleading. I am aware of a similar study done some years ago, where after 6 months of rigorous study, it was decided that price cuts of 20 cents a pound would expand the market for the product fourfold. What was overlooked was the fact that the product had unique property characteristics that made its demand inelastic, so that the company lost millions of dollars of sales revenues needlessly. Furthermore, the price cuts did not have the desired effect, because the product was already being used to the maximum extent based on these unique properties.

In order to safeguard against expensive strategic errors, we need to obtain, therefore, the following data:

1. If the product is new to the market, data on the size of the total market at various price levels.
2. If the product is not new to the market, data on the size of the market and the new product's share at various price levels.
3. The time needed for the product to reach the mature stage in the product life cycle at various price levels: this is necessary because time can affect costs, payback period and cash flow.
4. Distribution channels to be utilized, discounts, special promotional activities, and their related costs.
5. Reactions of competitiors, including possible retaliatory pricing, possible changes in their marketing strategies, and any new product development activity aimed at either upgrading their own products or at introducing new products of their own. The relative technical and marketing strength of possible competitors should also be assessed.
6. Total new investment costs at various production levels.

As the product is market tested, it becomes possible to set the final proposed prices in accordance with the findings obtained and after having decided on what pricing strategy to use.

Pricing Strategies

The landmark book in pricing is still the definitive *Pricing in Big Business,* published in 1958 (Kaplan, Dirlam and Lanzilotti), in which the authors explored the pricing strategies of a number of large corporations. However, a careful perusal of this admirable book sheds little light on pricing strategies for new products.

Briefly, two major strategy choices are frequently mentioned in new product pricing; skimming pricing and penetration pricing.

Skimming pricing, as its name implies, involves introducing a product at at a relatively high price. This works best when a product has unique characteristics that are relatively price inelastic. There is one important reason for introducing a product at a high price, rather than a lower one. It is easier to lower a price than to raise one, especially as competition begins to appear to chip away at the markets for the erstwhile new product. Skimming pricing is often successful for the following reasons:

1. Sales of the product are less likely to be price sensitive when first introduced rather than later.
2. Introducing a new product with a high price breaks up the market into segments with different price elasticities of demand. The high initial price skims the cream of the market, while subsequent price reductions enable the seller to locate the more price-sensitive market sectors.
3. A skimming policy is safer on the surface at least. The high price serves as a refusal price during the initial phases of market testing.

An obvious danger courted by skimming is that an "umbrella" may be provided for competitors.

Penetration pricing, the reverse of skimming pricing, uses low prices at the outset. This is often used when the product is price-volume sensitive, and when it is apparent that competition will appear very shortly after the product's introduction.

In choosing between two pricing strategy alternatives, a number of factors play a part:

1. The strength of the competition.
2. The degree to which the company can protect the uniqueness of its product, including patent protection.
3. The relative ease of entry of competitors. Low investment and operating costs naturally encourage competitors.
4. The forecast demand for the product. If the forecast market is small, this would act as a deterrent to new entrants, and vice versa.

The lowest possible price will obviously be equal to the costs plus the return on investment. However, it should be remembered that past costs incurred are not really germane to the price, because the costs are sunk. If the company's projected costs are higher than the estimates of what price is needed to introduce the product successfully, then the company obviously does not have a viable product, and the project should be abandoned.

The highest feasible price is, as mentioned earlier, related to the elasticity of demand for the product, which in turn would be a function of the uniqueness and degree of the product's performance characteristics. If, as is now most often the case, the product has only moderate or marginal performance advantage, the price would be not as high as would be the case with a truly unique product. This would also be true if there were relative ease of entry in terms of capital and technical requirements or if the product had inadequate patent protection.

Other comments on pricing strategies and approaches are noted below and have been treated in a number of excellent books and articles listed in the bibliography at the end of this chapter. However, for the sake of completeness, they are listed here. Whatever approach the price setter uses in the pricing decision, the guiding factor may be costs, the market, or competition.

When the guiding rationale is costs, the company may choose the strategy of:

1. Pricing to attain the established ROI standard
2. Pricing to obtain a fast payback of R&D costs
3. Pricing to cover full costs plus an industry mark-up rate the management may be stuck with
4. Pricing on conversion cost, with joint costing, or as a by-product
5. Pricing in relation to full product line and its profitability
6. Pricing on profit contribution to overhead, covering variable costs but not all constant costs

When the guiding rationale is the market, the strategy may entail:

1. Pricing to attain specific marketing goals of volume, share, niche, industry
2. Pricing to attract or encourage retailers, distributors, or agents
3. Pricing according to direct evaluation of customers' or markets' acceptance of the product
4. Pricing to increase the volume of other products, not for the product's profit (for example, safety razors at cost to sell the blades)
5. Price experimenting in test markets or in the whole market with subsequent adjustments (directly, through discounts, by product changes, and so on)

When the guiding rationale is competition, the strategy may be:

1. Calculating models above, at, and below their prices and adjusting the product's design, accessories, and so forth, to bring it within the desired gross margin if it is not
2. Adding a portion of the product's profit-rendering potential advantage (over the competitive product) to the competitor's price
3. Arbitrary or intuitive markup because of apparent product advantage
4. Pricing low for fast expansion or acceptance or to discourage competition: "penetration"

5. Pricing high for a luxury market, for quick profit, or for quick R&D pay-back (possibly anticipating lower prices later on entry of competion): "skimming"

Factors impinging on the price-setting decision are:

1. Forcasted profit cycle
2. Expected time advantage that the company can expect before entry by competitors
3. Impact of competition at that time
4. Estimated milestones when product improvements should be initiated
5. Milestones for price changes in the product during its life cycle
6. Effect of possible substitutes on the product's life cycle
7. Points of expected volume expansions and contractions
8. Effect of selling, advertising and promotion, campaigns and milestones for their implementation.

All of the above factors are inextricably intertwined, and it would be impossible to treat each separately. The following discussion, therefore, will examine the price-setting decison by discussing the role of costs.

Costs

As pointed out in Joel Dean's classic article "Pricing Policies for New Products," there are three types of costs that should be considered in marketing new products:

1. The company's projected costs in manufacturing the new product
2. The costs of present and future possible competitors
3. The cost of the product for the potential buyers

Company's Projected Costs

In projecting costs the company must keep in mind the life cycle of the new product and the fact that as the product moves through the life cycle it will probably be subject to a number of price revisions at different stages in the cycle. These will have an effect on the product's ultimate profitability, and if the company is not aware of this fact, it may end up losing money on the product over the long run. Also as the product moves out of the research stage and approaches the marketplace, it will be subject to price revision.

The producer whould establish a price that will enable him to select a final price that hopefully will maximize his return on investment.

The costs of the new producer can be divided into production costs and promotional costs. The production costs are those costs that will be added by making the product as opposed to not making it. Other costs will include investment costs R&D expenditures, product development costs, and all out-of-pocket expenditures that are associated with the launching of the new product.

Cost of Competitors

If there are already products in the marketplace that will compete with the new product, the company should attempt to estimate the price level beneath which the competitors will not go, that is, the *floor price.* A corollary of this statement is the need to estimate competitors' long-term and short-term incremental production costs.

The short-term incremental product cost would be of immediate relevance, whereas the long-term incremental production cost would be relevant in determining the effect of competitive products in forcing eventual reductions in the price of the new product.

If there are no present products that will compete with the new product, but competing products are either imminent or further down the road, the new producer should attempt to assess competitors' costs for such products. This is obviously a difficult task, because it means assessing the technical strengths of the would-be competitors and also their possible pricing strategy. As will be seen later, pricing strategy plays an important role in the ultimate profitability of a new product.

Finally, *buyer's costs* are important because, in the final analysis, if the product does not exhibit favorable price/performance characteristics, the would-be purchaser will evaporate.

In attempting to establish the probable price of the parameters of various prices of the new product, it is necessary for the producer to establish the value of the product to the potential buyers. This can be done in the following steps:

1. The end uses for the new product should be determined.
2. For each end-use area, determine the performance characteristics for the new product as compared with products already in the marketplace that are expected to compete with the new product.
3. For each end-use area, determine the price/performance characteristics of the new product as opposed to competitive products.
4. Estimate the sales volume anticipated for the new product in each end-use area.
5. Create a matrix of prices for each end use weighted by the sales volume.
6. Select the price that will maximize the sales revenue or profitability, depending on the company's objectives.

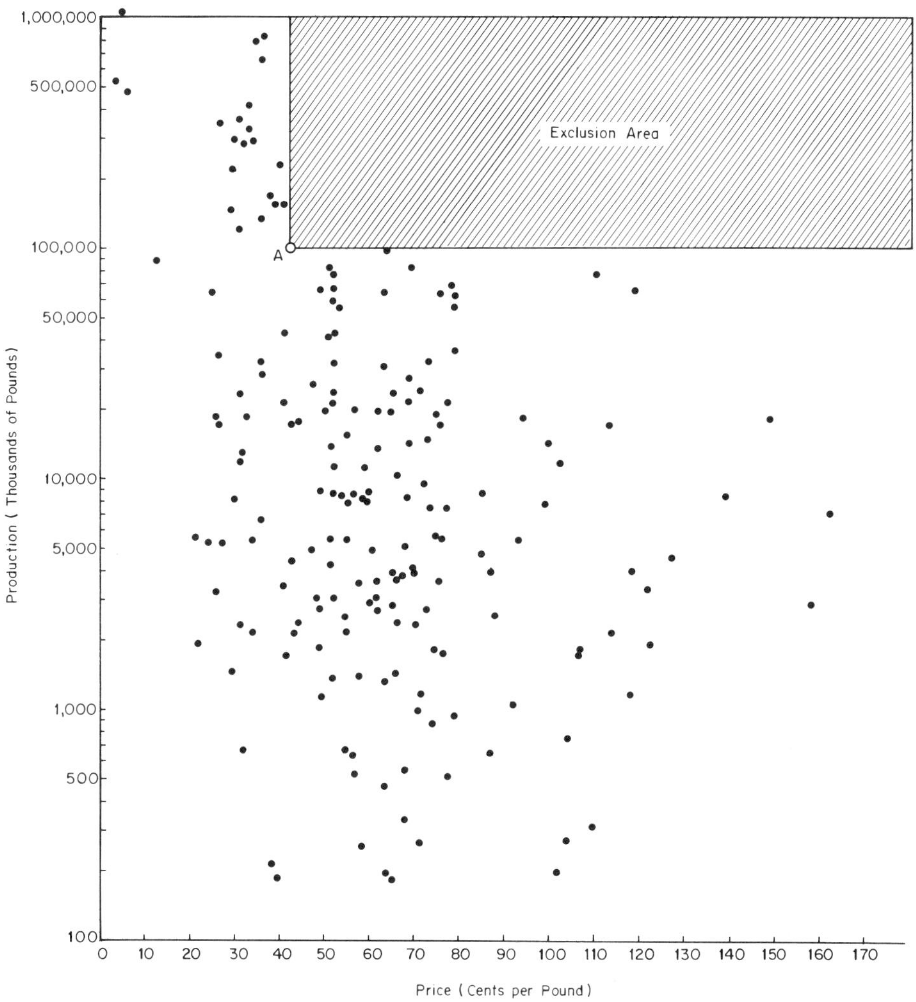

Figure 1. Exclusion chart for surface-active agent.

Exclusion Charts

A discussion of new product pricing would not be complete without a discussion of an invaluable tool to assist in the pricing process, the exclusion chart. The chart has been described in a number of seminars presented during the past 15 years by the staff members of Roger Williams Technical & Economic Services, Inc. The exclusion chart is simply a diagram on which have been plotted the price versus production volume figures for a large number of products for a similar end use.

The concept of the exclusion chart is quite simple. It merely states that if a product in a given product category is priced above a certain price level, the volume of sales of this product will be limited to a certain level, and below that price there is theoretically no limit to potential sales volume. Exclusion charts are strictly empirical in nature, and are based on actual sales and price data. In Figure 1, production volumes have been plotted against price for surface-active agents. These data were taken from the annual report on synthetic organic chemicals published by the U.S. International Trade Commission. Point A is the exclusion point, at a production level of 100,000 pounds and a price of 40 cents per pound. In other words, one could not hope to sell more than 100 million pounds of a surface-active agent if it were priced higher than 42 cents per pound.

These exclusion charts can be calculated over a number of years, so that a historical sense of perspective in a given category can be obtained.

Bibliography

Books

Brion, John M. *Corporate Marketing Planning.* New York: John Wiley, 1967.

Buell, Victor P., ed. *Handbook of Modern Marketing.* New York: McGraw-Hill, 1970.

Butrick, Frank M. *How To Develop New Products For Sale to Industry.* Englewood Cliffs, N.J.: Prentice-Hall, 1971.

Corey, E. Raymond. *Industrial Marketing: Cases and Concepts.* Englewood Cliffs, N.J.: Prentice-Hall, 1962.

——. *Industrial Marketing: Cases and Concepts,* 2nd ed. Englewood Cliffs, N.J.: Prentice-Hall, 1976, 1962.

Corley, H. M., ed. *Successful Commercial Chemical Development.* New York: John Wiley, 1954.

Giragosian, N. H., ed. *Chemical Marketing Research.* New York: Reinhold, 1967.

Hilton, Peter. *Handbook of New Product Development.* Englewood Cliffs, N.J.: Prentice-Hall, 1961.

Kaplan, A. D. H., Dirlam, Boel B., and Lanzillotti, Robert F. *Pricing in Big Business: A Case Approach.* Washington, D.C.: The Brookings Institution, 1958.

Lazo, Hector, and Corbin, Arnold. *Management in Marketing: Text and Cases.* New York: McGraw-Hill, 1961.

Marting, Elizabeth, ed. *Creative Pricing.* New York: American Management Association, 1968.

Nelson, J. R. *Marginal Cost Pricing in Practice.* Englewood Cliffs, N.J.: Prentice-Hall, 1964.

Oxenfeldt, Alfred R. *Pricing Strategies.* New York: AMACOM, 1975.

Perry, John H., ed. *Chemical Business Handbook.* New York: McGraw-Hill, 1954.

Pricing: The Critical Decision. Management Report No. 66, New York: American Management Association, 1961.

Risley, George. *Modern Industrial Marketing: A Decision-Making Approach.* New York: McGraw-Hill, 1972.

Spitz, Edward A., ed. *Product Planning.* Auerbach, 1972.

Steiner, George A. *Top Management Planning.* New York: Macmillan, 1969.

Stemp, Isay, ed. *Corporate Growth Strategies.* New York: American Management Association, 1970.

Talley, Walter J., Jr. *The Profitable Product: Its Planning, Launching, and Management.* Englewood Cliffs, N.J.: Prentice-Hall, 1965.

Articles and Speeches

Betley, Al J. "Contribution Pricing." *Management Accounting,* March 1973.

Biggs, Howard. "The Price of Research." *Management Today,* August 1972.

Bigson, C. J. "Selecting Pricing Strategies with Break-Even Analysis." *Cost & Management,* November-December 1972.

Brunk, Max E. "Pricing Consumer Items." *Management,* April 1973.

Burck, Gilbert. "The Myths and Realities of Corporate Pricing." *Fortune,* April 1972.

Dhalla, N. K. "The Art of Product Pricing." *Management Review,* June 1964.

Dean, Joel. "Pricing Policies for New Products." *Harvard Business Review,* November-December 1976.

Eaton, William W. "Patent Problem: Who Owns the Rights?" *Harvard Business Review,* November-December 1967, p. 101.

Fox, Harold W. "Pricing and Planning." *Managerial Planning,* March-April 1972.

Hakala, Donald R. "Pricing for Profit Planning." *Managerial Planning,* January-February 1973.

Isakson, Hans R., and Maurizi, Alex R. "The Consumer Economics of Unit Pricing." *Journal of Marketing Research,* August 1973.

Lanzillotti, Robert F. "Pricing Objectives in Large Companies." *American Economic Review,* December 1958.

Lanzilloti, Robert F. "Pricing of Chemical Products." *Chem. Eng. News,* April 30, 1962.

Levin, Jules. "The Strategy of Pricing." *Industrial Canada,* April 1973.

Monroe, Kent B., and LaPlaca, Peter J. "What Are the Benefits of Unit Pricing?" *Journal of Marketing,* July 1972.

Nimer, Daniel A. "Developing A Strategy for Pricing." *Management Review,* November 1971.

Oxenfeldt, Alfred R. "A Decision-Making Structure for Price Decisions." *Journal of Marketing,* January 1973.

Roberge, Michael D. "Pricing for Government Contractors." *Management Accounting,* June 1973.

Rogoff, Donald L., and Lynn, Robert A. "Methods vs. Objectives in Pricing Policy." *Management Adviser,* March-April 1972.

Seglin, L. "How to Price New Products." *Chemical Engineering,* September 16, 1963.

Shapiro, Benson P. "Price Reliance: Existence and Sources." *Journal of Marketing Research,* August 1973.

Simons, Leon. "Market Economics and Pricing." *Management Decision,* Winter 1972.

Thompson, G. Clark, and MacDonald, Morgan B. "Pricing New Products." *The Conference Board Record,* January 1964.

Walker, Arleigh W. "How to Price Industrial Products." *Harvard Business Review,* September-October 1967.

Williams, Roger. "The Exclusion Chart Revisited." *Chemtech,* October 1973.

Zabel, Herman W. "Use of the Exclusion Chart to Forecast Price." Speech before the Division of Chemical Marketing and Economics of the American Chemical Society, April 11, 1967.

Zabel, Herman W., and Marchitto, Michael. "The Exclusion Chart." Speech before the Division of Chemical Marketing and Economics of the American Chemical Society, April 6, 1959.

10
THE COMMUNICATIONS PROGRAM

Philip Gisser
Foster D. Snell Division
Booz-Allen & Hamilton Inc.
Florham Park, New Jersey

Communications, or advertising as it is commonly called, is normally one of the key elements of a new-product launching. In most situations, it is the public exposure of the product in a communications program that marks the transition from the controlled market development phase of a new-product program to the commericial introduction phase.

Although it is normally an important step, the scope of the communications program varies a great deal from situation to situation. This variation is related largely to the number of people who have to be reached by the product message and the reliability of identification of the people who will be influential. If the audience is small and clearly defined—as it would be if the new product is a control element for atomic power plants or a system useful to credit card companies—the communications program may consist of nothing more than personal letters and contacts. Where the market is broad, as it is for a new office copier or a new cleansing agent, the communications program becomes the pivotal element of the introduction program—and one of the major cost considerations as well. For a new specialty material or scientific instrument whose market may be small, but where the potential buyers are not identifiable, the communications strategy is likely to become the crucial factor in product success or failure.

The significance of communications to most products falls somewhere between the extremes. Communications may well be a significant element in product success, and the cost of communications may well be a significant factor in product economics, but for most industrial products it is not a dominating factor. Whether it turns out to be dominating for a specific new-product situation depends primarily on the two factors cited above—the number of people who have to be reached and the reliability with which they can be identified.

Because communications is an important element of most new product programs, it is important for project managers to understand the various tools of communication and to know how to make them work for them in their overall program. It is not important that they know the detailed tools of the trade. There are many experienced people and organizations that can provide project managers with qualified judgments and services in this area—usually the advertising department of the company or an advertising or sales promotion agency. What is important is that project mangers have an appreciation for the effects of the basic strategic decisions that they will make in this area and that they know enough about the techniques applied so that they can communicate efficiently with the advertising professionals who will serve them in this area. It is to that end that this chapter is directed.

Communications Techniques

Although people frequently apply the term "advertising" to the communications field as a whole, it does create a problem in semantics, because "advertising" is used within the communications field to refer to one particular technique—the use of paid space to carry the product message. Although that particular technique is generally the dominant one in terms of budget requirements, it is not always the most important in an industrial new product situation.

The broader term "communication," or "marketing communications," as it is often referred to, is a far more accurate description of the function involved. Under it is included all of the techniques wherein communications directed to multiple recipients are sent from a single source. Whereas a sales call or a business letter falls into the class of "direct communications," the form letter and the packaged sales presentation fall into the catagory of replicated "marketing communications."

There are a number of techniques that fit under this classification. The obvious ones are space advertising, television advertising, publicity, direct mail, and literature. But there are a variety of other techniques, too, such as trade shows, demonstrations, seminars, movies, and inquiry handling, each of which will be described shortly. Each technique has its balance of advantages and disadvantages, as well as an individual cost structure, which makes for different patterns of cost-effectiveness for each.

Rarely are all the techniques used for a single situation. Rather, some mix of the available tools is generally selected to provide the maximum cost-effectiveness of an overall program. A key element of communications strategy, to be discussed later in this chapter, is the proper selection of the mix of techniques to be utilized for a specific new product introduction. To make the optimum judgment, however, each of the various techniques must be understood. A brief description of the fundamental characteristics of each technique follows.

Space and Electronic Advertising

Space advertising is so well understood and thought of as the key element in communications that most people understand its characteristics well. Its concept is simple and clear. The medium, whether a business publication, newspaper, or television station, applies editorial or entertainment techniques to gain the attention of an audience with certain specific characteristics on a consistent basis. It does this by catering to the interest of the group desired in editorial pages or programming content. This audience is then made available to firms who want to reach that audience with their specific messages—for a fee, of course. Because a specific publication carries the messages of a large number of companies at the same time, it can do this at a far lower cost than would be involved if an advertiser were reaching the same audience on its own.

Countering this economy, though, is the problem that normally the publication, in trying to serve many advertisers, defines its audience in somewhat broader terms than the individual advertiser does. So part of the savings is eaten up because the advertiser message necessarily goes to people who are not prospective customers, a part of the circulation generally referred to as "waste circulation." The closer the media circulation matches the desired audience, the better the media is from a cost-effectiveness point of view. The higher the waste, the poorer the economics of space media looks in relation to other communications alternatives.

The activities involved in use of paid advertising space generally fall into two basic groupings—advertising preparation and media activities. The preparation activities relate to the preparation of the advertising message in the most effective form and its production into the mechanical form needed by the media to integrate it into its production. Media activities relate to the selection of the media and the arrangements that must be made to get the advertising into the selected publications.

Advertising preparation, though by far the smaller element of cost, is probably the most important from a performance point of view. During this phase of development, the advertising message is selected, its technique of presentation is developed, and it is put into the form required by the media. In the case of magazine or newspaper space advertising, the end product is a "plate" or "film" ready to be incorporated into the graphic arts processes involved in printing the publication. In the case of electronic media, the finished piece may be a film, videotape, or audio tape ready for airing.

The many steps involved in translating the message into these physical forms are too numerous and too complex for detailed discussion in this chapter. In general, advertising agencies have developed expertise in handing these steps, so the many decisions and steps involved are best left to the professionals in such organizations. The areas where the new product management team should involve itself in detail are agency selection and compensation, strategy development, and technical approvals. If these steps are handled properly, the rest of the activities can normally be delegated to competent professionals.

The other main element of space advertising is media-related activities—primarily media selection. No matter what the product, there will be a large number of publications that in one way or another seem to make sense as an advertising investment. In the case of industrial products, the choices are usually somewhat limited, but in consumer areas they are almost infinite. But even the limitations in the industrial or business areas offer too many choices. If a product goes to three industries, and each industry has six significant publications serving it, plus an equal number of publications serving the functions defined as key purchasing influences, the number is already far larger than an introduction budget will normally tolerate. The problem then becomes one of selection from the wide variety of publications that offer themselves as candidates.

The basis for selection generally involves balancing two factors—reach and frequency. Reach is a measure of the percent of the desired audience that will be exposed to the message; frequency relates to the number of times such exposure can take place. In consumer media, much of this data has been quantified by survey techniques. Media specialists in advertising agencies have been trained to analyze such data in ways that are meaningful to the particular objectives desired.

Although the same two parameters, reach and frequency, are equally important in business and industrial markets, the data are generally in a form that makes the development of realistic reach and frequency data far less scientific, and more in need of judgmental inputs. Basic circulation data, usually audited for accuracy by an independent audit bureau, are available from the various publishers and are published in the *Business Publications Edition of Standard Rate and Data Service.* Usually these data are broken down into Standard Industrial Classification (SIC) and job function classifications, though often these classifications are not completely comparable between publications. A major effort is underway to develop such comparability, and it has been achieved in a number of market areas.

Although two publications serving the same industry may be compared with some degree of reliability, the problem exists of comparing two unlike publications that serve the same people. For example, if the primary target is electronic purchasing agents, the question arises as to whether it is more efficient to reach them through electronics publications or purchasing publications.

Although each field may be comparable within itself, there is little objective guidance for selection between fields. And practically all of the data available refers to the primary recipients of a publication. The two to four people who generally see a magazine after the initial recipient are generally ignored.

These are only a few examples of the difficulties that surround the media selection process. Fortunately the professionals and the advertising agencies do have the necessary experience to mix the many elements of objective data with the equally important nonobjective elements to make the necessary judgments as to how to achieve maximum effectiveness for the dollars allocated to advertising media.

For them to make these judgments, however, they must be given the best possible definition of the audience sought and a clear indication of the budget limitations within which the media program must fall. It is generally far more productive for project managers to spend their time working out these parameters with the advertising professionals than to spend it in attempting to learn the intricacies of media selection.

There is one area of media selection guidance that the project manager can provide more efficiently than the media professionals. In the preintroduction stage, a market research study is often made in the effort to develop the parameters that will guide later market targeting. Such studies, whether made by personal interview, telephone, or mail, generally seek to determine who the key buying influences are and where they are located. If the habits of the magazine's readers can be determined at the same time, these data can provide primary guidance for the media selection step that is to come later. It generally costs little or nothing to add a question on publication readership to an existing questionnaire, and its value can be very substantial in guiding the later advertising program toward optimal media decision. The only reason it is not done more often is that few people think about its value during the early information-gathering stages.

There are many other elements of space advertising decision making that may be significant to the success of the new product. Aside from media selection, decisions must be made on creative approach, the size of the ad, use of color, frequency within each of the publications, headlines, art, and so on. Project managers must separate these clearly into "strategic" and "tactical" decision areas and should become involved only in the strategic ones. The tactical decisions can then be fully delegated to advertising professionals. These strategic decision areas are discussed in more detail later in this chapter.

Publicity

Publicity is the use of editorial segments of the media to carry the desired communications message. This technique is especially applicable in new product introductions, because new product news is of high interest to readers, and it is, therefore, the kind of thing that editors want to include in their publications to maintain the interest of their readers. Thus publicity often becomes the most efficient communications vehicle for new product introduction in terms of results per dollar invested.

It is not quite so efficient that it warrants the description that is commonly applied—"Free advertising." The space is free, it is true, if we earn the right to it. But to earn that right, some investment must be made in getting the necessary information and in putting it into the form that the editors need. Without these efforts, the opportunity may be lost, and once lost it is lost forever.

The communications results of publicity are not really comparable to those obtained by means of space advertising. Whereas advertising pages deliver the message in exactly the manner the advertiser has selected with the emphasis on the points that are felt to be most important, publicity passes through the hands of a middleman, the editor. It is often changed, its direction modified, and it can even be given a negative implication before it is delivered to the audience. And whereas advertising can be repeated over and over again to reach those who did not read the magazine the first time or who missed the particular item, publicity is used just once and is then forgotten. On the positive side, though, information that apears in the editorial pages is generally better read and given higher credibility by the audience—far more than space advertising, which is so frequently discounted to some degree by the reader. And publicity can appear in many more publications than an advertiser can normally afford to place his advertising; thus it reaches a far larger audience, even if much of it is a fringe, rather than a key, audience.

Professional publicity people use a wide variety of techniques to maximize their publicity reach. They include news releases, press conferences, and editor contact.

New Release

This is a simple written statement of the news story, composed in such a way that enough information is presented from which the editor can build a story. It states all the details in a way that minimizes the opportunity for error. In a new product introduction, the initial press release usually announces the availability of the product, describes it, and tells what its advantages, disadvantages, and likely applications will be. This may be followed at a later date by other releases announcing successful applications, modifications, key people appointments, and so on.

Press Conference

The press conference is another commonly used technique in the new product area. Here a number of editors are called together for a meeting in which the new product is presented and frequently demonstrated as well. The news release plays a part in this, for it generally becomes the heart of the *press kit,* the compilation of materials that serves as source data for the stories that the editors who attend will likely prepare. Other elements of the press kit might include photographs, names of key people attending, literature, and background material on the problem that the new product solves. Sometimes several news releases are included on different aspects of the story.

Editor Contact

Another publicity technique is simple one-to-one contact with editors to work out a story line that meets the editorial requirements of the publication, as well as the communications objectives of the company.

Publicity is without question one of the biggest communications opportunities in the new products area. It is a sensitive technique, though, whose results are unpredictable. For that reason it is best left to professionals for handling. Fortunately, such professionals are readily available, either in specialized public relations agencies or perhaps attached to the advertising agencies that serve the broader communications areas.

Direct Mail

Direct mail, or multiple mailings of a single communication, can play a very important role in new product introduction, especially if the new product is destined for a relatively narrow and clearly defined industrial market. The choice between use of space advertising or direct mail to reach a given audience can be largely a matter of economics. It is always cheaper to reach a reader via space advertising than with a mailing. The cost of reaching a good prospect, however, may be very different, depending upon the waste circulation of the publication involved and the accuracy with which a mailing list can be drawn. If the smallest publication in the target area has a circulation of 30,000, but the product has only 1000 potential users, it is probably more economical to reach the 1000 by direct mail than by space advertising—assuming that the 1000 can be identified with some degree of accuracy.

But direct cost comparisons are not the only criteria for selection of the type of media. Often things can be done with direct mail that cannot be done with space advertising. Delivery of a complex brochure is one, and distribution of a sample is another. It is easier, too, to put a response card in the hands of the prospect through direct mail than through space advertising.

The techniques involved in using direct mail break down into two major areas: preparation of the mailing package and development of the mailing list.

Package Preparation

Preparation of the mailing package is similar to the production end of space advertising. The message must be crystallized and reduced to paper—an effort that can best be undertaken by communications professionals. For direct mail, though, a far wider range of creative execution is possible, for there are fewer limitations on such things as size, thickness, number of pages, form, number of elements, and so on. Perhaps the direct mail program should be nothing more than a personal letter written to a number of individuals at the same time. Or perhaps sophisticated creativity is needed simply to get the mailing piece through to the recipient without having it thrown away unopened or screened out by the recipient's secretary.

List Development

List development is a different problem, and one that must be considered before a decision is made to use the direct mail medium. There is simply no point in relying on direct mail as a communications vehicle unless one can count on having a list that includes the right people. There are many sources of list information—publishers, list brokers, and mailing list formulators. The best source information as to what lists are commercially available is the *Direct Mail List Edition of Standard Rate and Data Service.*

If the mailing piece contemplated is relatively inexpensive it is often best to use a purchased list, even if it contains a certain degree of waste. But if the mailing piece is expensive, it is usually worthwhile to build a list. This might be done by starting with a purchased list and screening out the nonprospects or by finding names from the directories and building up the list. Another technique is to develop a list of known prospect companies and then obtain individual names by phoning the companies involved. Or if only one piece is to be mailed, a mailing might be made to certain titles within each company with only a small loss of effectiveness from a personalized list mailing. For a program involving multiple mailings, though, it is far better to use individual names.

Literature

Literature is a basic need of any new product introduction, whether it takes the form of a simple data sheet, a product brochure, or an educational treatise. It is needed to handle inquiries. It is needed by sales personnel to give to the prospects they call on. It is needed by prospects to remind them of what they heard from the salesperson and to help them communicate with other buying influentials within their company.

Although there is rarely a serious question of whether or not to include literature in the communications mix, there is always a serious question as to what kind should be produced and what portion of the available resources should be allocated to this effort. For some product situations, a technical data sheet may be all that is required. For others, perhaps an instruction sheet is sufficient. For products that require a great deal of concept sell, however, the literature requirements may well demand both an educational treatise on the subject and a product brochure.

For most new technical products, an eight to twenty-four page brochure seems to fulfill the requirements. The preparation of such a brochure is inevitably difficult and time-consuming, for unlike some of the other techniques, it cannot be done most efficiently by outside professionals, at least in the early development stages. It needs the time and attention of the people who are to provide the basic information. It needs the time and attention of the people who will write the copy and prepare it for production. It needs the time and attention of the project manager to make sure it is consistent with the remainder of the program. And no matter who does it, it needs time for the physical processes of mechanical production, printing, and binding. Too often, the people on whom the literature development depends in its early phases are tied up on other aspects of the new product introduction and fail to give this the priority it deserves—at least until the project is faced with the problem of getting underway with an essential part of the communications program incomplete. More than any other part of the program, the literature project must get underway early, or it can become the bottleneck that threatens the effectiveness of the entire new product introduction.

Exhibits

Exhibits play an important role in many new product introductions in the business and industrial area. This is because so many industries have exhibits that are acknowledged as the place where all the key members of a given industry gather to communicate their new developments and to learn about the new developments of others. Often the exhibit occurs once a year, and not infrequently on a 2-, 3-, or even 5-year cycle. Thus the timing of the product introduction often gets tied closely to the exhibit schedule.

But timing is not the only reason why exhibits are so important in new product introduction strategy. They are important because they provide an opportunity to communicate to a lot of people within a short period of time, a situation that matches the objectives of most introduction programs. And exhibits are unique among the communications techniques in that they offer immediate feedback from prospects—consistent with a key objective of most introduction programs to get a quick reading of the level of acceptance and the degree of interest among prospects. In exhibits we have the ready-made response mechanism, for the product is exposed directly to prospects, and, if handled properly, they have an immediate opportunity to feed back their reaction.

Exhibits are especially important for products such as machinery, which may be too big to move to the prospect, but which can be set up and even operated in an exhibit hall. Other types of products that lend themselves to the exhibit media are auxiliary equipment, control systems, and materials destined to be handled on mechanical equipment. Although exhibits can be used for products such as portable equipment, additives, services, and small parts with some success, one must remember that for these there are alternate communications techniques available that may prove to be far less costly.

With all these advantages for the right product, exhibits create a real problem for most new product introduction programs. The space usually has to be contracted for long before the product release date can be known with any degree of certainty. And if there is some slippage in the latter stages of product development, the already-planned exhibit may become the factor that leads to an unfortunate decision to launch the product before it is really ready.

Even if the timing does work out, the introduction of a product at a show often has other disadvantages. A lot of new products are likely to be introduced at the show, and the result is a dilution of the publicity on all. And because the publicity does not normally appear until after the show, potential customers who attended the show might easily miss the new product in the confusion of the multiple offerings and the strident atmosphere characteristic of industrial exhibits.

If the development timetable can support it, the most effective way to utilize an industry show for product introduction is to release the publicity in time for it to appear before the show begins. In this way, people whose interest has been whetted by the publicity are likely to search out the new product at the exhibit.

Demonstrations and Seminars

Another communications option is the privately sponsored demonstration that is unrelated to an industry show. This is most applicable for capital equipment that can best be shown by demonstration. This technique generally combines the use of other mass communications techniques to draw the audience to the demonstration and then a well-organized demonstration that makes the most of the opportunity established. Where a regional roll-out of the product is selected as a market introduction strategy, a series of regional demonstrations often makes sense.

A seminar approach is not very different, except that the audience is gathered for educational purposes rather than for demonstration—though the two can be combined in some cases. This is generally most applicable if the concept of using the product must be communicated before the product itself can be accepted. The seminar, therefore, overcomes the concept barrier and paves the way for the sales of the product. This technique is not likely to be useful if the new product is similar to something already on the market, but it can be very effective for highly innovative products or services.

Movies

The movie, or videotape, is a logical alternative to the demonstration technique, especially if the demonstration will have to be repeated many times. By translating the demonstration into a movie, it can be repeated anywhere—to large groups gathered for the purpose or to individuals contacted by the sales personnel. The movie may be even more effective than a demonstration if the innovative action element of the product is hidden to the eye, or if things move too fast to be seen.

The production of a movie, however, is usually an expensive and time-consuming activity. And for a new product, both the expense and manpower required must be justified on a short-term basis. Too often, such a production is rationalized on the basis of its long-term usefulness. Most new product offerings, however, often change in direction after introduction, and thus a fixed presentation, such as a movie, may become obsolete long before its payout time is complete.

Inquiry Handling

Many readers may be surprised to find inquiry handling listed among the communications techniques. Yet it meets the criteria in every way. A package of inquiry-handling and follow-up materials is developed and then directed to the people who inquired about the product. In many ways it is similar to direct mail, except that whereas in direct mail the list is developed and the mailing sent out at a single point in time, in inquiry handling the list is developed over a period of time, and the mailing is sent out as each name is identified.

The lists developed by inquiries are one of the important elements of new product introduction. Such a list contains the names of people who have identified themselves as interested in the product. This provides not only a sound marketing base for follow-up but a sound sample for further market research as well, which is especially useful in evaluating the future course of the project.

In general, the follow-up of inquiries should involve some prescreening before personal sales calls are made. A letter or telephone screening can usually reduce the number of active prospects down to 20 to 30 percent of the total inquiries received. These are the ones that should be followed up by the sales force.

Communications Strategy Is the Responsibility of Project Management

New product management involves intelligent decision making. And with communications normally an important segment of the new product launch activities, communications decision making becomes an important element of product success or failure.

In communications, as in the other important areas, there are two major areas of decision making–strategic and tactical. Tactical decisions, such as selection of magazines to carry the advertising, art technique, copy and headline treatment, or reproduction technique, are best left to the professionals in the communications field. Strategic decisions, on the other hand, must be made by those responsible for moving the new product into the marketplace–with the counsel, of course, of the communications specialists. Some of the major areas of strategic decision making are detailed in the following sections.

Objectives

The role of communications in the overall mix is a fundamental strategic decision. The objective must be clearly defined before the communications tacticians can begin to approach their work.

The primary objective of the overall communications program may be to solicit inquiries or to make people aware of the product or simply to invite people to a demonstration or a combination of several objectives. The selection of the primary objective and the secondary ones is the most important strategic decision that must be made.

The objective of any communications program, whether for a new product introduction or not, relates to the effect of the communications upon the recipient of the message. Thus all objectives fall into one or more of three possible classifications

1. Action objective–get people to do something
2. Educational objective–get people to learn something
3. Attitudinal objective–get people to think differently about something

Action and educational objectives are most commonly applicable to new product introductions. Attitudinal objectives are far more common in the case of established products and company image campaigns. There are times, though, even involving new products, when attitudinal objectives play a key role, so all three types will be discussed in detail.

Action Objectives

Action objectives are easiest to visualize. An obvious action objective is direct sales. For products with a wide audience appeal and relatively low trial risk, an attempt to solicit orders by communications techniques (usually space advertising or direct mail) can often provide key information for test marketing as well as some revenues to help support the overall effort, whether or not the direct solicitation route is visualized as a future distribution technique. Products such as personal, household, or entertainment items can often be sold in this way while the distribution patterns are being built. And low-priced industrial products, such as chemical specialties, can be direct marketed, too.

Another common action objective in industrial new product introduction is self-identification of the prospect. The key purpose of the communications, in this case, is to obtain inquiries for the new product. This provides the names of people who have a need that is somewhere close to that which the product is designed to fulfill. These names can then serve as a basis for direct sales contact or for solicitation by a further direct mail program. The names solicited also serve as a basis for further market research aimed at pinpointing the effectiveness of the product presentation at a later stage.

Another action objective that is often applicable to new industrial materials is feedback of information. Often the potential areas of use of a new material are not known. Here, the main purpose of the advertising may be to get information on the market segments that demonstrate the greatest interest in the product.

Educational Objectives

Educational objectives are most important for new products. In this class of objectives, the end result desired from a communications program is a change in knowledge level of the audience. For a new product, the precommunications knowledge level of the product's existence is normally close to zero. The key objective, then, is to make the potential buyers of the product aware of its existence. And with that, of course, goes the need to define it in terms of the potential benefit it offers and its differences in relation to other products in the same area. Only when there is a high level of knowledge of the product availability and its characteristics can the next steps in product acceptance begin to take place. If there is a failure in achieving a level of awareness, the other product introduction factors become meaningless.

There are many other education objectives that might be applicable in a given new product situation. Making people aware of a specific product attribute is an example. "The new product will last five times as long." "The new product has more capacity." "The new product is cheaper to operate."

And in highly innovative technical products where the new product acceptance requires an understanding of a new way of doing things, the technology may well have to be taught to the audience before the product concept can be accepted. This was the case, for example, in the introduction of each new control technology, electronics, fluidics, and pneumatic moving part logic.

Similarly, where the acceptance of a new product depends on getting customers to make a change in their operating technique, the advantage of making the change must be communicated before the product itself can be accepted. Thus the educational objective must relate more to the operating technique than to the product itself. The introduction of prestressed concrete and the equipment involved are typical examples in this area. Another example is the introduction of powdered polyethelyene as a raw material for rotational molding–a process that had little commercial significance until the development of powdered polyethylene. In this case, education on what could be done with rotational molding and what was involved in the process became the prime objective of the communications program.

Attitudinal Objectives

Attitude-change objectives generally involve creating an image or changing an impression in the mind of the prospective buyer. Objectives within this class are common in the advertising of existing products and services and the companies that supply them. We've all seen the "Banks aren't really stuffy" ads, the "We're really a diversified company" ads, and the "It doesn't have to taste bad to work" ads–all of them attempting to change an attitude in a segment of the audience as a prerequisite to making sales.

In new product situations, attitude-change objectives are generally less important, if only because the education objectives tend to predominate. There are some exceptions, though, which arise principally where some established attitude or prejudice is so firmly established among the prospects that it presents a significant obstacle to sales. For example, so many computer companies have come and gone that buyers are fearful of the stability of any but the top names in the field. A company that introduces a new product into this area, especially one that is central to the computer installation rather than peripheral, must establish an image of business stability if its product is to succeed. Similarly, a mechanical device that is likely to require service must establish an image of service availability as part of a new product introduction program.

Image objectives are applicable also when the image is in itself a part of the product offering. This is usually limited to new products that are at some extreme of the positioning scale. "The most luxurious." "The cheapest." "The most sophisticated." And so on. In these cases the key product attribute is image, and the image and communications objectives blend together.

There are other elements of communications objectives, beyond the desired effect on the audience, that must be considered as well. They are definition of the audience, time frame for meeting the objectives, and the method by which the success or failure of the communications will be judged.

Definition of the audience is all important in the framing of advertising strategy and media tactics. This definition establishes the target—the universe of people among whom the desired effect is to take place. The way of defining audience differs significantly, however, between industrial and consumer markets.

In industrial markets the audience definition is usually framed in terms of the product or service produced by the company and the job function of the people who have purchasing influence on the product. The nature of the company is often designated by the Standard Industrial Classification of the products made by the company. This is an extremely useful way of describing the market, because most publications break down their circulations in this same manner, and the relative efficiencies of reaching the audiences of the publications can therefore be put on an objective basis.

Frequently, though, other ways of defining an audience can be more meaningful. High-technology products might find their buying influences only among people who have been trained in a particular technology, and so professional training may be a more meaningful way of defining that audience. If we desire to reach chemical engineers or physicists or electronics engineers or analytical chemists independently and not through their companies, there are publications that define their audiences in those terms.

Another audience definition may be by "equipment owned," rather than by industry or technology. If the new product is a plastic resin for blow molding, the target audience is limited to people with that equipment. Although this kind of definition helps the strategist to focus more closely on a target, it is often hard to apply this procedure in media selection, unless the publications under consideration define their audiences by the same criteria. Direct mail lists of owners of specific types of equipment can occasionally be found.

For consumer products the key elements of audience definition are the demographic characteristics of the audience—age, sex, economic status, marital status, educational level, and so on. It is especially useful to define the audience in these terms because the major consumer media defines its audiences in this way. Thus it becomes possible to apply objective criteria to the media selection process.

One element of audience definition that applies to both industrial and consumer products is geographic location. This is more commonly a segment of consumer audience definition for two reasons—consumer products are more frequently introduced into local markets, and consumer media are frequently local or regional by nature; if national, they sometimes offer local or regional circulation options. Most industrial products are launched on a national basis, and some on a worldwide basis. With more and more frequency, though, industrial products are being introduced on a regional basis. Frequently this relates to either the service requirement or distributor organizations that are required locally. Industrial publications are, therefore, beginning to offer regional editions to satisfy that need.

This ever-present requirement to define the audience in absolutely clear terms as a prerequisite to designing an effective communications programs points to the need for effective market research during the preintroduction period, with at least one of its objectives to provide that audience definition. If the nature of the product is such that a clear definition is not practical until the product is introduced, the only alternative is to communicate to an audience broad enough to include all the probable buying influences and to use the results of the first phase of communications to develop the audience definition. New sensing devices and new materials such as exotic alloys, nonmetallics, and plastics often need this kind of handling. But we have to keep in mind the fact that communications costs relate to the number of people reached—not to the number of potential customers included in the group, and so this approach can prove very expensive.

Communications Parameters

There are a certain number of communications parameters that must also be fixed before the advertising professionals can be expected to develop an optimum program that fills the communications needs of the new product. The key ones other than timing and cost, which are discussed later, are quality level and specific restraints.

Quality level, a term that refers to the quality of production of the communications elements, may or may not be a prime parameter, depending on the situation. In general, the quality parameter should be consistent with the position of the product in the marketplace. Especially in consumer products, the higher the product position is in terms of quality, the higher must be the minimum standard that all of the elements of the communications program must meet. If your product is the "Cadillac" or "Tiffany" of its area, all ads, literature, presentations, commercials, and so on must utilize the highest quality production techniques, and must be done in the best possible taste. Though this means a higher dollar commitment, anything less would leave the communications image at odds with the product position and be a total waste of money, if not a negative factor.

On the other hand, a product that is positioned near the bottom of the quality scale can use the cheapest of reproduction techniques and invest its communications funds in obtaining higher reach. For some products, high quality production could hurt rather than help.

The new product that is being introduced by a large corporation has quality standards superimposed upon it above and beyond that of the product. The corporation has its overall image to protect, and these standards must supersede the requirements of the product. Thus a product that has no real quality image requirement of its own might still have a requirement that all its communications meet a given quality standard.

Whatever the minimum, the important thing at this stage is to make sure that the communications professionals know whether or not there are any quality limitations, and if there are, exactly what they are. Thus the program they devise can be consistent with the needs, and its costs can be estimated realistically.

Specific restraints arise in a relatively small number of situations. Yet where they are applicable, they must be considered and communicated to those involved in communications planning if their plans are to meet the product communications needs.

In medical products or insecticides, for example, legally required warnings must appear on specific types of communications. For politically sensitive consumer products such as alcoholic beverages, a large number of legal and industry-wide voluntary restraints must be taken into consideration. Product and marketing strategy sometimes impose specific restraints too. For a new color photographic process or graphics arts material, for example, the restraint that all printed materials be in color is highly logical. A paper product should logically use the product as the base for printed material describing the product. And the project manager for the "biggest machine in the industry" may decree that all printed material be over standard size. There may or may not be such restraints on your specific product, but if there are, they must be taken into consideration at the time when budgets are being developed.

Timing

It often seems as if there are only two choices of the timing for a public introduction of a new product—too early and too late. If we time it too early, we may have technical or production problems ahead of us that would require withdrawing the product—and there is nothing harder than the "reintroduction" of a new product. But if we wait too long, not only are revenues delayed, but competition may jump in ahead and take advantage of all the image and communications benefits that go to the first one who introduces the product.

For communications guidance, a decision must be made as to when to first introduce the product to the public—a time when prices must be fixed, service facilities deployed, inventory maintained, and the reputation of both the company and the product is laid on the line. This is a key decision, and one that must be made by the manager of the new product, to serve as a guidepost for all the supporting services involved as well as the communications professionals.

The expectation of results from the communications program must be placed in a time frame as well. It is not enough simply to set an objective of soliciting 2000 inquiries or distributing 5000 samples—a time limit must be set on when this is to be accomplished.

The choice between media may well depend on the timing and time frame selected. Exhibits may loom high on the communications technique list if the introduction is to be made at one time—but low at another. If one objective is to make 60 percent of the audience aware of the product within a year, publicity may be assigned the major part of the communications task. If the objective is to do the same thing within a month, direct mail, daily newspapers, and electronic media may be the only real choices.

Positioning

The term "positioning" means different things to different people, but fundamentally it relates to the relationship of the new product to others in the market place as perceived by the potential buyer. Many of the key positioning decisions are made during development of the product, and in the strategic decisions made in areas other than communications. The basic features designed into a new product go far toward positioning it against its competitors. And price is also always a key positioning element. A decision to design a product specifically for one segment of a market is a positioning decision. When these decisions have already been made, it is the role of communications to pass them on to prospects, so that the new product is recognized by the audience to have the characteristics designed into it.

But communicating market segmentation is only part of the positioning assistance that can be expected from communications. Communications can work from the preconceptions of the audience and built-in differentials in ways that traditional market segmentation can not. For example, the product, whose acceptance involves more risk than its better established competitors, can position itself for the adventurous people among its prospects. For the market that is dominated by a single supplier, the logical position might be "equivalent in performance to the leader, but less costly."

Budget

The budget for the communications activities is another key strategic decision to be made. It is up to the project manager to allocate his resources properly, and to decide where the available funds can be invested for maximum result. Because communications is often a substantial cost factor, the amount allocated is a key decision.

The budget allocation cannot be set arbitrarily, however. It must be set in conjunction with the objectives. In general, the cost of communications is directly related to a number of factors, including the number of people to be reached, the level of uncertainty as to whom to reach, the number of communications objectives, and the complexity of the communications objectives. The availability of media that are concentrated on the desired audience is another factor that has a heavy bearing on budget requirements as well.

It is important to involve the communications specialists in the budget area, so as to get the best possible inputs on the trade-offs involved, before the final budget is decided. One good way of approaching the problem is to have them develop several preliminary budgets at several levels, each related to the communications objectives it will support. Often a maximum budget, a minimum budget, and one or two intermediate levels will provide the guidance necessary for decision making.

Media Mix

Although the choice of which magazine to use or which direct mail list to use clearly falls into the tactical decision area, the choice of types of communication media is a strategic one. In a sense, it is a decision as to how to allocate the communications budget described above.

The media mix decision can be a key one to new product success. A decision to emphasize direct mail rather than space advertising implies that there is a clean-cut definition of the prospect audience. If this is true, a lot more communications for the money might be the result. If, however, it turns out that the judgment as to who will buy the product was in error, the product might fail simply because the right people were never reached by a too narrow direct mail list.

For guidance as to how much to allocate to each communications technique, the project manager must again rely on inputs and judgments from the advertising professionals. Here, though, it is harder to get unbiased evaluations, because people tend to recommend the use of the type of media with which they are most familiar. Imagine, for example, the difference in media mix recommendations you would receive from an advertising agency, a sales promotion specialist, a direct mail specialist, a public relations agency, and an exhibit builder. For this reason, the direction being recommended should be probed carefully by the decision maker until he or she feels comfortable that a logical media mix is planned.

The Launch Must Be Properly Sequenced

It is most important that the communications steps of the new product launch be handled in proper sequence. Failure to do this results in lost opportunities and reduction of the cost-effectiveness of the precious dollars devoted to communications. For example, the opportunity for broad coverage of an announcement news release will be lost if the space advertising comes out too early and scoops the editor in his or her efforts. And much of the interest elicited from the market during the first wave of communications will be dissipated if completion of the literature is delayed. And the cooperation of the sales staff can never be obtained if it turns out that they are the last ones to learn about the new product.

To assure that the communications are organized and planned in such a way that an optimum effectiveness timetable is achieved, the "critical path," or PERT, technique provides an ideal planning medium. Although the two terms have some difference in meaning for highly sophisticated planners, they both involve planning and sequencing the detailed elements of a complex program in terms of the various steps that have to be accomplished.

A PERT diagram consists of a group of blocks each representing a specific and identifiable point in the progress of a project. The blocks are arranged on paper with connecting arrows indicating which events come before and which come after each step. This provides a clear picture that shows what parts of the project can proceed concurrently and what parts must be accomplished in sequence. From such a diagram, the manager can easily determine the shortest possible time in which the entire job can be done and where the potential bottlenecks lie. The following is an example of how the network anaysis technique actually was used in a new product situation.

The new product was "Ultrathene," U.S. Industrial Chemical's trade name for a new family of ethylene vinyl acetate copolymers, a series of plastics with many of the performance properties of rubber. The material had been in the market development stage for over a year, and several companies had developed applications that were close to commercialization. At least two other companies were developing products of the same type at the time.

One segment of the audience was clearly defined as plastics and rubber processors. A second part was less clearly defined as industrial designers and others with materials responsibility in a wide variety of user industries. The sales force had had no previous contact with rubber processors or with the wide variety of end users to whom this product would be of interest. It was clear that mass communication techniques must be used quickly and efficiently if competition was to be preempted.

The first determination in using PERT was the desired sequence for completion of each element of the program. The key to the timetable became the date of the press conference, which was dictated by the closing dates of some of the major publications involved.

The information had to be presented to the sales staff before the press conference date, and the presentation to the sales staff had to include both the news release and the literature. This fixed the date by which the literature and release had to be completed. The press conference date also fixed the time by which the physical arrangements for the press conference had to be completed. This included such things as developing a list of invitees and making arrangements for catering of refreshments.

Because many of the magazines serving this particular market are monthlies, the company knew it would be at least 1 month before the publicity articles resulting from the press conference would be published. Knowing that the publications prefer not to have an announcement ad appear in the same issue as their news articles, this fixed the earliest date for appearance of the ad at some 45 working days after the press conference. And knowing that this would bring in inquiries, it directed that the complete inquiry-handling system be in operation about 3 days after the direct mail piece was sent out.

The next big problem was prepartion of the many elements of the program within the timetable. Certain activities had to be completed before other elements of the program could get underway. For example, basic technical information on the product had to be received before literature, the sales manual, or the news release could be prepared. Case histories were needed for the news release and the sales manual. Activities such as publicity and preparation of data sheets could proceed without specific budget authorization, but procedures called for budget approval before the space advertising, direct mail, and the literature could be prepared.

Once the time limitations at the end of the program were established and the starting points established at the beginning, it was relatively easy to list the key activities in the preparation of each of the segments. For example in the space advertising segment, the key steps were the following:

1. Ad program proposal complete
2. Budget submitted
3. Budget approved
4. Ad program approved
5. Ad copy-layout complete
6. Ad copy-layout approved
7. Ad mechanicals complete
8. Ad mechanicals approved
9. Ad electros to publication
10. Ad appears

Similar steps were involved in every other segment of the communications program.

All of the events, and their preceding and succeeding events, reduce to a single complex-looking diagram, as in Figure 1. Each block represents a clearly defined event, and each connecting line represents an activity. The numbers on the lines represent the number of working days estimated to be needed for that particular activity.

The next step was to identify the critical path—the series of steps that takes the longest time. It was apparent in this situation that there were two critical paths (shown by the heavy lines). One critical path involved preparation of the brochure. The other involved prepartion of the direct mail piece. As a matter of fact it was apparent that there were not enough days available to complete either program.

By calling attention to these time limitations, the diagram made its greatest contribution. The brochure was given highest priority and pushed hard in the early stages, so that it was completed in less time than the original estimate. Preparation of the direct mail piece could not be speeded up, however, because it involved an outside supplier who had to mold samples of the product, and so a news release with a note attached was substituted, the timetable was maintained, and the more complex mail piece became the second mailer in the program.

It was especially interesting to note that preparation of the ad, the project usually thought of first in a situation like this, could be delayed until after the other elements were completed. This knowledge permitted the company to turn maximum attention to preparation of the other materials. The diagram clarified something else too. Though the sales manual and the literature were needed at the same time, the literature had a much longer production time. Thus, a technical writer was able to finish writing the literature and then turn his attention to the sales manual while the literature was being produced.

It was apparent that follow-up of the inquiries would strain the capabilities of the field sales force. Arrangements were therefore made to telephone screen all inquiries so that only the active prospects would be called to the sales staff's attention. Difficulties were also anticipated in filling sample requests from the company's plants, so it was arranged to have a contract packager repackage the material into smaller containers and handle the mailing of the sample requests.

Along with each sample went a letter describing the material, a postcard enabling prospects to acknowledge receipt, and an evaluation form the user could fill out to describe his experience during evaluation. These were invaluable in eliminating unnecessary calls on the part of the already busy sales staff.

Of course a lot of auxiliary forms were required to mechanize the inquiry and sample handling systems, but the planning system pinpointed the time requirements, and they were completed on schedule.

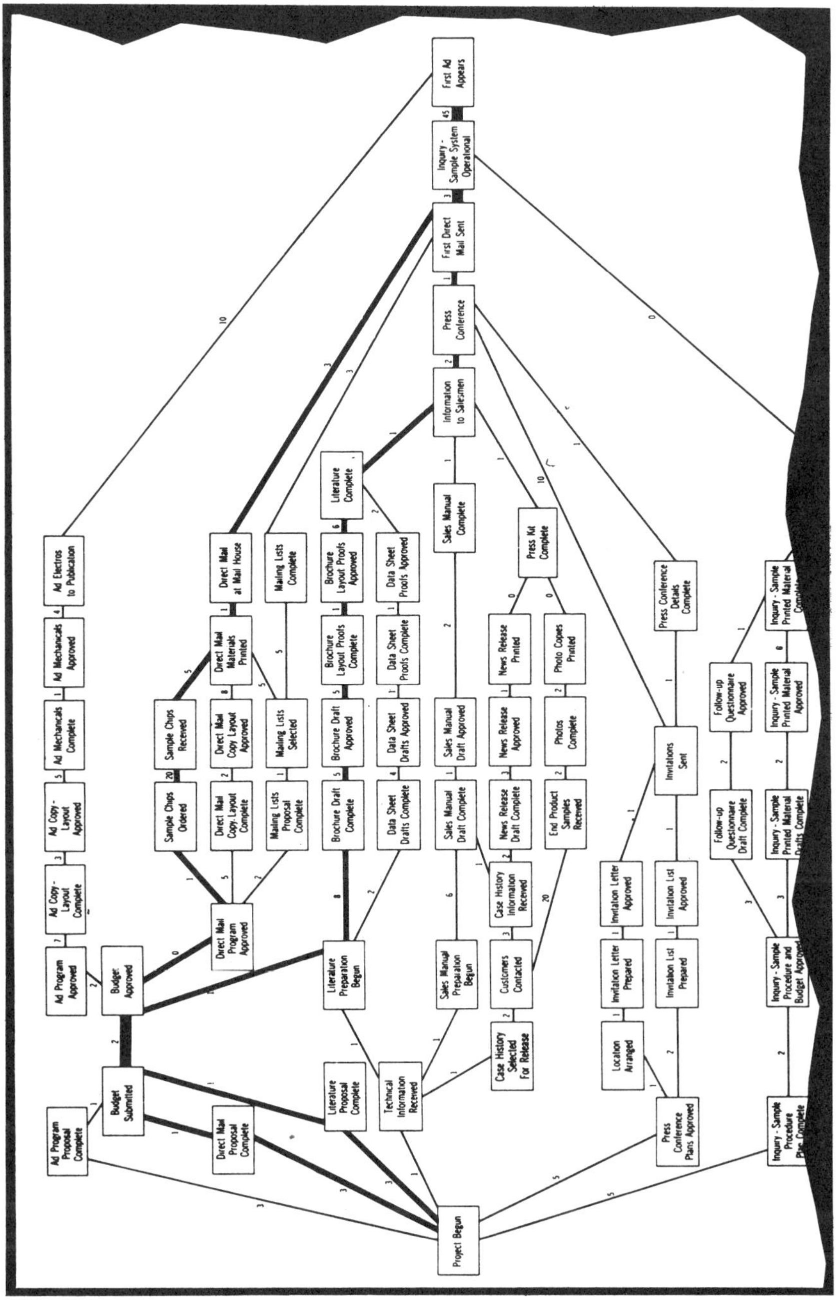

Figure 1. Heavy black lines indicate the two critical paths—the series of steps requiring the most time—in this new product introduction program.

Once the planning was complete, and the various parts of the project were in preparation, the chart helped to provide control. As steps were completed, the blocks were colored in, so that it served as a visual control board.

The program, undertaken and completed within a 2-month period, was an unqualified success. Six months after introduction, the inquiry count was up to 4000. All were handled, screened, and either relegated to a "dead file" or turned over to the sales staff within 6 weeks after receipt. And although sales figures are confidential, the product moved out of the "introduction" phase and into its "growth" phase in a period of less than 6 months.

Although the program put a greatly increased burden on both sales people and technical service people, neither group was at any time strained to the point where they had to pass up good prospects. And with such a clear-cut lead established, largely by virtue of the PERT planning and sequencing, U.S.I. Chemical dominated this particular market for many years.

Bibliography

Alexander, Ralph S., et al. *Industrial Marketing,* 3rd ed. Homewood, Ill.: Richard D. Irwin, 1967.

Business Publication Rates and Data. Monthly, Standard Rate and Data Service, Inc., Skokie, Ill.

Coram, Terry C., and Hill, Roy W., eds. *New Ideas in Industrial Marketing.* New York: International Publications Service, 1970.

Corey, E. Raymond. *Industrial Marketing: Cases and Concepts,* 2nd ed. Englewood Cliffs, N.J.: Prentice-Hall, 1976.

Dening, James, ed. *Marketing Industrial Goods.* Boston: Cahners Books, 1968.

Direct Mail List Rates and Data. Semiannual, Standard Rate and Data Service, Inc., Slokie, Ill.

Dodge, B. F. *Industrial Marketing.* New York: McGraw-Hill, 1970.

Dominguez, George S. *Product Management.* New York: American Management Association, 1971.

Fisher, Lawrence. *Industrial Marketing.* New York: Petrocelli Books, 1970.

Gisser, Philip. *Launching the New Industrial Product.* New York: American Management Association, 1972.

Kelly, Richard J. *The Advertising Budget; Preparation, Administration, and Control.* New York: Association of National Advertisers, 1967.

Lucas, Darrell B., and Britt, S. H. *Measuring Advertising Effectiveness.* New York: McGraw-Hill, 1963.

Messner, Frederick R. *Industrial Advertising; Planning, Creating, Evaluating and Merchandising It More Effectively.* New York: McGraw-Hill, 1963.

Risley, George. *Modern Industrial Marketing.* New York: McGraw-Hill, 1972.

Schiller, Robert, ed. *Market and Media Evaluation.* New York: Macmillan, 1968.

Wilson, A. *Marketing of Professional Services.* New York: McGraw-Hill, 1972.

11
EXTERNAL SOURCES OF COMMERCIAL DEVELOPMENT ASSISTANCE

Newman H. Giragosian
Delphi Marketing Services, Inc.
New York, New York

Needless to say, successful commercial development cannot be accomplished in a vacuum. On many occasions, all but the largest companies need outside assistance in developing new products and businesses. Outside assistance takes several forms, but can be classified as follows:

1. Consultants
2. Industrial research institutes
3. Universities and philanthropic organizations
4. Trade and professional associations
5. Government sources
6. Literature sources

Consultants

The problem of whether or not to use consultants in assisting a company's commercial development activity has always been a thorny one. However, with the proliferating use of consultants in industry it seems feasible to discuss the pros and cons of using consultants and how best to select consultants and make use of their services.

Consulting services are available for every phase of industrial development; they cover not only those functions associated with commercial development but also research, process development, plant design, location, and construction. However, in this section only consultants specializing in various phases of commercial development will be discussed.

Consultants are frequently individuals working alone or with a small staff; in some cases they may be fairly large organizations with many departments and specialized facilities. Naturally the larger consulting organizations are equipped to render more complete services than the individual consultant. However, the consultant on a retainer basis becomes more familiar with the specific problems of his or her client and is more like an employee than an outsider. A chemical consultant is often employed for one specific problem and upon its solution is no longer required; in many other cases retention is on an annual or longer-term basis, and the consultant is considered to be a part-time employee. When a consultant has rendered satisfactory service for a client over a period of time, there is a natural tendency to select him or her for more assignments. No standard practice exists for these arrangements.

Advantages of Using Consultants

There are a number of advantages to be gained by using consultants in commercial development. Some of these are anonymity, cost saving, objectivity, improved contacts, and imaginative ideas.

Anonymity

Consultants can serve as a means of anonymously obtaining information not generally available to a client without disclosing the purpose for which it is desired; this is particularly helpful in surveying markets for a new product when it is desirable that the company's interest not be identified. Also, in certain types of projects there is the possibility of obtaining biased results if the company being interviewed knows the name of the interested company.

Cost Savings

Very importantly, consultants can be retained for short-term problems without incurring the obligations or expense involved in setting up a new department in the company or employing additional personnel. The consultant is paid on a fee basis, and withholdings for income tax, social security, and other employee benefits are avoided. This is an important saving in these days of constantly rising costs for fringe benefits.

Objectivity

A consultant often provides an objective, fresh approach to problems, which is difficult for an in-house group to maintain when confronted with everyday problems. A good consultant, by contributing the fresh, detached viewpoint of a friendly visitor, and knowing the approach of others to similar problems and being familiar with both company objectives and capacities, can be of great assistance to the client. A consultant provides a good inexpensive means of obtaining a qualified unibased opinion to check conclusions reached within the organization.

Improved Contacts

Consultants usually have a wide circle of acquaintances in industry, which often allows them entrée to many avenues of information that would be closed to an in-house group. These generally include potential customers, raw-material suppliers, and others of interest to a client.

Imaginative Ideas

A consultant often works with several firms in different fields of activity. Because of this he or she is frequently able to make suggestions or ask questions that would not occur to company employees of more limited industrial experience. The consultant is usually aggressive, alert, imaginative, and a good source of new ideas; if he or she did not have these qualities he or she probably would not long survive as an independent consultant.

There are a number of situations in commercial development where the services of an outside consultant may be helpful, such as:

A large company with a completely staffed commercial development group needs advice on an area of their industry with which they are not familiar.

A producer of a certain line of products that are currently being sold to one or two industries may want to adapt its products to another industry where items with similar properties are being used.

A company may wish to conduct product opportunity studies in industries or countries with which they may not be familiar.

A consultant is valuable in many different situations of these types. It is usually uneconomical to employ a full-time expert, even if one can be found, when a consultant having specialized knowledge along certain lines can quickly furnish the information at a relatively small cost.

Selecting the Consultant

It is essential that when a company decides to use an outside consultant it selects the proper one. The selection process, however, is not as difficult as it may appear at first blush, in spite of the proliferation of consultants, most of whom are qualified but many of whom are "waiting between jobs."

There are several means by which consultants can be selected and evaluated.

1. Previous experience is by far the best method, but it might also be limiting. There is nothing like a satisfied customer, and a happy client will tend to gravitate toward the same consultant for the next job, regardless of his or her qualifications for that particular project.
2. Recommendations of other companies who utilize consultants. The reputation of a consultant spreads by word of mouth, and a consultant who is in demand has often achieved this status by providing quality work.
3. Sometimes, although not often, there are only one or two organizations that can provide the necessary facilities for a certain project.
4. Preliminary discussions with a number of consultants often can quickly determine their qualifications for a certain project.

The problem of selection, though not difficult, can be laced with pitfalls. For instance, the client should not place too much credence on the physical facilities of the consultant or on the size of his or her operation. Consultants vary in size of their operations from a single-person "shop," through loosely banded groups, to larger, more formally organized and structured companies. Size, however, is no criterion of the qualifications of consultants.

How does the would-be user of an outside consultant develop his or her leads? The easiest and probably the best is to ask his or her peers in other companies for the names of qualified consultants.

The best sources of leads are people who have already made use of consultants' services or who know of the experience of companies who have. Obvious contacts are commercial development executives, trade association executives, trade journal editors, and so forth.

Directories and rosters of professional associations are other sources of leads. Some of the directories used are:

Consultants and Consulting Organizations Directory, published by Gale Research Company, Detroit, Michigan.

Directory of Membership and Services, published by the Association of Management Consulting Engineers, 347 Madison Avenue, New York City.

Directory, Association of Management Consultants, published by the Association of Management Consultants, 947 Old York Road, Abington, Pennsylvania

International Directory of Marketing Research Houses and Services, published by New York Chapter, Inc., American Marketing Association, 60 East 42nd Street, New York City.

European Guide to Industrial Marketing Consultancy, published by the Industrial Marketing Research Association, Litchfield, England.

Selection Criteria

The Conference Board has done a number of studies on the use of consultants and suitable selection criteria. These surveys indicate that in choosing a consultant, a company should consider the following:

1. The specific qualifications of the individuals who will be performing the actual work. Many cases of disappointment on the part of the client occur when consultants whom the client has not met are brought in to do the work with those who were involved in selling the assignment taking a minor role. It is important for the client to know not only who will work on the assignment but the role that each person will play.
2. The degree to which the consulting firm, through discussions and its written proposal, appears to understand the client's problem.
3. The breadth of talent available within the consulting firm that could be made available to solve related problems.
4. The extent to which the consultants can and will make themselves available to assist in interpretation and implementation guidance.
5. The estimated charges for the assignment.

Steps in Selecting a Consultant

Although methods may vary from company to company and according to the magnitude of the job to be done, the following steps have been recommended as a general systematic approach to selecting a consultant:

1. Prepare a statement outlining the nature, general scope, and purposes of the assignment to be undertaken prior to contacting prospective consultants. Such a statement will assist the client in selecting for interviews the consultant(s) most suitable for the assignment. It will also help potential consultants to present their qualifications and capabilities to undertake the work.
2. Consider the general qualifications of a number of management consultants who appear to be capable of meeting the requirements of the assignment and select for interview one or more (probably not more than three) who appear to be the best qualified for the proposed assignment. Reference checks on general reputation should be a normal part of the client's preliminary search for consulting help.
3. Preliminary discussions should be held with each of the consultants selected to discuss the project and their approach to it, and they should be asked to submit proposals.
4. The client should then study the proposals in terms of their understanding of the problem, approach, probable benefits, cost, and the particular experience and ability of each consultant to meet the requirements of the engagement. In weighing this experience, care should be taken to consider the qualifications of the personnel who will actually be working on the project.
5. References of those consultants being seriously considered should be carefully checked.
6. The final selection should be based on a careful deliberation of the above factors. Final negotiations should follow the selction of the consultant.

7. Compensation for professional services of management consultants may be established by a variety of methods. Among the more common modes of compensation are the following:
   a. Per diem or hourly fee
   b. Bracket quotation
   c. The lump sum or fixed amount
   d. The retainer fee
8. Penalty clauses, performance bonds, or other "incentive" features are unnecessary provisions in a professional service contract or agreement. Such devices often provoke a reduction in quality of service to the client and place emphasis on less important aspects of the contract or agreement.

As pointed out earlier, consultants vary considerably in size. The important thing, naturally, is to select the consultant best qualified to meet the needs of the assignment. The company's problems will probably receive more attention from the principals of a small firm, but employing a so-called "name firm" may carry greater weight. Then again, there are a number of qualified individual practitioners who could do an excellent job. The size question is one that the client must decide.

As pointed out in the excellent monograph *How to Get the Best Results from Management Consultants,* when selecting a consultant, it is well to bear in mind that the heart of the consulting process is the analysis and solution of problems by independent and objective professional people who enter the field largely because they like to solve problems, particularly new ones. They find a new problem more challenging than one they have solved many times before. They will work harder and with deeper interest on a new problem than an old one.

Many clients feel more confident in employing consultants who have specific knowledge and experience in their kind of business and industry. However, it is important to recognize the fact that professional consultants have developed the ability to analyze and solve problems without necessarily being conversant with all the minute details of the business.

Some executives have built-in prejudices that limit their freedom of choice. They shy away from younger consultants who have obviously not had a great deal of operating experience, failing to realize that skill in sizing up a situation does not necessarily coincide with the ability to do the job itself.

Others are reluctant to deal with newly established firms. If these firms are made up of persons who have had a successful record of industrial or other experience in the areas in which they are offering services, they should be given careful consideration because they will be anxious to establish a good reputation for themselves. The fact remains that although reputation is a useful guide, it must continue to be earned.

Still others look primarily at fees, a practice which can be misleading for many reasons. Companies sometimes receive proposals from several different consulting firms, with widely varying estimates of costs for the project. The fee should be commensurate with the nature of services to be performed and the responsibility to be assumed by the consultant. There should be no significant variation in the cost of work that is comparable in both quantity and quality. When several reputable consultants submit widely varying estimates of cost for an apparently identical assignment, it is certain that they are actually proposing to perform widely varying quantities or qualities of work. The approach of each consultant to the problem and its solution must be taken into account when considering the fee. In consultant services, as in anything else, the important factor is the result. A good job is worth the cost; a poor one is a loss, no matter how attractive the price may have been. Professional consultants are not inexpensive, and they should not be.

Occasionally companies call in many consultants and obtain proposals from all of them before making a selection. This practice has no particular advantages for either the client or the consultant. It is not particularly beneficial for the client to have to evaluate a large number of proposals, and the consultants whose proposals are rejected will have gone to considerable expense to prepare them. Naturally, in any competitive bidding there will be a winner and some losers, but it is wasteful to have an unnecessarily large number of losers. This practice also has a tendency to encourage overly optimistic estimates of costs and savings and other conduct of doubtful professional caliber and almost always puts the conscientious consultant at a disadvantage.

Obviously the best course is to select several consultants who are highly qualified to meet the requirements of the study, discuss it with them, and ask them for proposals.

There are a number of advantages in a continuing relationship with a well-qualified consulting firm. A consultant who has worked for a client over a period of time is familiar with the company and does not have to become oriented before tackling the next problem. In addition, he or she is able to perform economically. Obviously, his or her interest in the welfare of such a client is greater than that of a firm that might be called in for a single specific study.

It is also advisable to make a reference check on the qualifications of the consultant. The following questions are useful to keep in mind:

1. How long has the firm been in business?
2. What are the backgrounds of the principals?
3. How much time will the principals of the firm spend on the assignment?
4. Is the staff composed of people who view their jobs as a profession and are, therefore, reasonably permanent?
5. Do you think you would enjoy working with this firm?

6. What kind of clients has it served?
7. How much of its business is repeat business?
8. What is its general reputation in the business community?
9. Has the firm had experience that is applicable to the problem?
10. Is it the practice of the firm to make a proposal in writing, clearly defining the objectives of the assignment and the work to be done?
11. Does the firm provide an estimate of fees in advance?
12. Are the estimated charges reasonable, considering the difficulty of the problem and the time required to solve it?

Appraisal of the Consultant's Work

After the project is finished, the client company should then formally appraise the results of the consultant's work. Too often this is not done in a formal fashion, so that as time goes by, memories fade, the personnel involved with the project leave the company for other positions or retire, and the unsatisfactory consultant may be given the opportunity to obtain another job at the same company.

The following questions should be used as reference checks:

1. What was the nature of the work done by the consultant?
2. Did the professional staff assigned to the project demonstrate professional competence, objectivity, and integrity?
3. Did they work constructively with client personnel?
4. Did one or more principals spend sufficient time in supervising the work?
5. Were the solutions developed complete, timely, practical, and suited to the client's specific needs?
6. Was the suggested course of action the most effective and economical from the client's point of view?
7. Was the work accomplished within reasonable time and fee limits?
8. What is the overall evaluation by client executives of the value of the consulting engagement?
9. Would the client retain the consultant again if it had need for the consultant's kind of professional services?

Industrial Research Institutes

There are a number of reasons why companies use industrial research institutes for commercial development assistance:

1. To obtain facilities and abilities not presently available within the company, especially when the company is a small-to-medium-sized organization

having a limited research department and technical staff. The institute provides trained personnel and a broad knowledge of industry.

2. To supplement its own facilities. Large corporations with highly developed research departments often need a broader approach and viewpoint than can be obtained with their personnel.
3. To obtain temporary help. The commercial development of one or a small group of products is often short term; when markets have been developed the need for the commercial development effort often disappears. It may not be advisable for an established research department to invest in special equipment or specialized personnel for a short-term project. Furthermore, when the project is finished many of the personnel are released or absorbed into jobs for which they are not well suited. Research institutes have a continuing influx of new projects and are geared to handle projects of varying duration, so that they can readily take on problems that are short term.

In almost all cases, the research institutes operate on a nonprofit basis. The cost of their services is about the same as that of a comparable effort in one's own laboratories.

Research is undertaken on a fellowship basis or on a project basis. By contractual agreement, institutes usually commit themselves to do research for the benefit of the client on a carefully defined subject and at a stated rate of effort. Either the actual budget or a fee ceiling is made a part of the agreement. In practically all cases, patentable results become the property of the client. The institute pledges secrecy to the extent desired by the client. In practice, the selection of an institute is quite similar to the selection of a consultant.

Universities and Philanthropic Organizations

As would be expected, universities are useful as adjuncts to the ongoing commercial development effort of a company. The personnel of universities are specialists in many fields and supply expertise that most individual companies can hardly match.

Assistance can be of several kinds, such as performance of specialized tests in a number of basic research areas, detection of harmful trace impurities in a new product, and toxicological studies.

Although university personnel lack commercial backgrounds, their broad knowledge of various fields of science can be valuable in commercial development. For instance, let us suppose that a company has defined a new market and the characteristics of a hypothetical new product to meet the needs of that market. The equipment and personnel of the university can then be utilized to develop products to fit that need; these products can then be evaluated by the company's market development personnel for commercial feasibility.

Philanthropic research institutes have narrower and more specific objectives than those of universities. They are usually supported by private funds and are often dedicated to long-range study of specific areas, such as sugar research, plant, and major problems. Institutes can be helpful in a planned program of continuing study on a narrow subject, such as studying physiological and toxicological properties of new products.

Trade and Professional Associations

Trade and professional associations are extremely valuable sources of outside aid. Some of them provide data on the industries they serve, such as the Society of the Plastics Industry, the National Lubricating Grease Institute, and the Chlorine Institute, among others. A more important function, however, is that of providing a starting point for further investigation. Association directors can and often do supply valuable inputs, such as suggesting knowledgeable people to contact and other sources of information to explore.

Listings of trade and technical associations are found in the *Encyclopedia of Associations* (Gale Publishing Company, Chicago, Illinois) and *National Trade and Professional Associations of the United States and Labor Unions* (Columbia Books, Washington, D.C.).

When seeking assistance from trade associations, it should be remembered that members of trade associations are corporations or companies, whereas members of technical associations are individuals.

Examples of trade associations are the Manufacturing Chemists' Association and the Synthetic Organic Chemicals Manufacturers Association; examples of technical associations are the Commercial Development Association and the Chemical Marketing Research Association.

Because the members of trade associations are companies, such associations exist solely for the benefit of their members. Usually all members of a given association are in the same industry and are competitors. It is not logical to expect an association to be a source of information about the operations or affairs of an individual company. The information available from this source is of a broad, industry-wide nature, and the technical assistance that may be available will pertain to uses of the member companies' products rather than to production technology.

The second point is that there are many different kinds of trade associations and no set pattern of activities. Some are concerned solely with promoting the use of a specific product. A larger proportion of them have more diversified interests and activities. The common points that hold members of a trade association together may be any of the following:

The raw material used

Processes employed

The character of the products

The importance of having their industry recognized

A combination of several or all of these factors

Associations have been formed by participants in almost every form of commercial venture. According to figures released by the Commerce Department on national, regional, and local associations, over 35 percent represent the retail trades; about 20 percent represent the service trades; and about 15 percent represent the manufacturing industry. Membership may be based on function (retailers, distributors, or manufacturers) or on product (people concerned with all aspects of the citrus fruit industry). There are also federated trade associations, composed of associations that belong to a "parent" organization, such as the National Forest Products Association. There are associations of associations and individuals both, such as the American Standards Institute. The American Society of Association Executives estimates that the total number of U.S. trade associations, their local chapters, and independent local or regional groups is 40,000. One person or company can sometimes be eligible for membership in several dozen. It has been estimated that there are now more than 2500 national trade associations in the United States and more than 2000 professional societies.

Various studies that have been carried out on the activities of associations indicate the following activities in order of importance:

1. Issue publications of various types
2. Disseminate information to the general public on their industry or products
3. Gather statistics for dissemination of members
4. Sponsor and/or carry out product research
5. Carry out market research
6. Provide a library service

Other activities relate to business practices, education and training, public health, safety, standardization, and public relations.

Those activities most germane to the commercial development practitioner revolve around obtaining information on statistics and economic trends and the publications issued by the various associations. Almost all trade associations compile and disseminate statistics to their members, and about half of them collect and compile original statistical information that is not available from any other source. Some associations not only complile data related to their industry but also appraise economic trends in their own and other industries. Other acitivities of interest include the sponsoring and carrying out of scientific research, the results of which are made available to their members.

Association research is usually directed at developing basic information about either raw materials or a product in which all the members of the industry would be interested. Research on a product is generally designed to discover new uses and increase demand. These results are made freely available to the public. The research on a raw material is for the benefit of the participating companies, and the results are not generally made available to the public. Although information concerning specific production, processes, and operating methods used by the member companies is not available from a trade association, the processes and techniques involving member companies' products are widely disseminated.

Government Sources

Needless to say, it is impossible to carry out any kind of commercial development effort without coming into contact with the government at state and federal levels. At the same time, product development can be aided by having access to the informational and technical facilities of the proper governmental agency. The various states exercise their own laws and regulations and are sources of information on products manufactured or used within their boundaries, but the federal government is the principal regulatory agency to be considered in any development program.

It is extremely helpful to have a general working knowledge of the various federal and state agencies, particularly those that one wishes to contact. An awareness of their work in various fields can be of valuable assistance in obtaining specialized information. Because of the multitude of government departments, bureaus, and agencies, it is not always easy to locate the proper authority, and the task can indeed be formidable. It is indeed frustrating and costly to spend half a day trying to locate the appropriate individual in the government with whom to establish contact. Directories of the various agencies showing locations and telephone numbers of personnel are obsolete almost as soon as they are printed.

It is difficult to offer advice on how best to proceed in locating a specific individual. Broadly speaking, there are several guidelines one can follow:

1. Consult the *United States Government Manual,* which is revised each year. This is designed as a basic introduction to some of the significant information outlets provided to aid access to government information soucres.
2. *The Federal Statistical Directory,* issued sporadically, lists, by agency and appropriate organizational units, the names and locations of key persons engaged in statistical programs.

3. Most of the key agencies publish directories showing the location and telephone number of all their personnel. As mentioned, these tend to become obsolete very quickly because of retirements and transfers, but they can serve as a useful starting point.

Of major interest to the commercial development practitioner are the data provided by the government agencies, some of which will be discussed further in this section. A guide to the data available is published in the *Monthly Catalog of Government Publications,* and the monthly listing of the publications published by the Bureau of the Census is also useful.

Several new agencies have come into existence that have not only had a tremendous impact on industry but also supply a vast amount of information not available heretofore. The activities of the Occupational Safety Hazard Administration (OSHA), the Environmental Protection Agency (EPA), the Federal Energy Administration (FED), and the Consumer Product Safety Commission have opened up new areas. In addition, the National Technical Information Service (NTIS) has correlated and coordinated the reports and other information available from the federal government. The International Section of the Commerce Department commissions studies that later become available to the general public at modest cost.

The information section of the Administrative Procedure Act (5 U.S.C. 552) was amended in 1966 by what has become popularly known as the "Freedom of Information Act." This act became effective on July 4, 1967, and was itself amended by Public Law 93-502, (88 Stat. 1561) effective November 21, 1974.

The Freedom of Information Act provides for making information held by federal agencies available to the public unless it comes within one of the specific categories of matters exempt from public disclosure. The legislative history of the act (particularly the recent amendments) makes it clear that the primary purpose was to make information maintained by the executive branch of the federal government more available to the public. At the same time, the act recognized that records that cannot be disclosed without impairing rights of privacy or important government operations must be protected from disclosure.

Virtually all agencies of the executive branch of the federal government have issued regulations to implement the Freedom of Information Act. These regulations inform the public where certain types of information may be readily obtained, how other information may be obtained on request, and what internal agency appeals are available if a member of the public is refused requested information. To locate specific agency regulations pertaining to freedom of information, consult the General Index of the Code of Federal Regulations under "Information availability." (Many agencies give the location of reading rooms for public inspection of documents under the heading "Soruces of Information.")

The commercial development practitioner should be familiar with four basic publications: *Synthetic Organic Chemicals, The Census of Manufacturers, Minerals Yearbook,* and *Agrictural Statistics.*

Synthetic Organic Chemicals, U.S. Production and Sales (International Trade Commission)

Probably the most valuable reference source, and the one that researchers have often called the "bible" of source data, is *Synthetic Organic Chemicals, U.S. Production and Sales,* published annually by the International Trade Commission (ITC). Data on the production and sale of about 2000 individual chemicals are supplied together with gross data on a total of about 6000 chemicals. This publication records the history of the synthetic organic chemical industry from World War I to the present.

In order to release information as promptly as possible for use by the chemical industry, the ITC issues a preliminary report for each major category months prior to the annual report. In addition, monthly reports are issued for a selected list of organic chemicals and plastics.

U.S. Department of Commerce

The U.S. Department of Commerce offers a vast variety of publications. As a guide to the publications issued, the *Business Service Checklist,* issued weekly by the U.S. Department of Commerce, lists reports published by that department and other government agencies.

The *Marketing Information Guide* is an annotated listing of selected current reports (governmental and private) of interest in marketing studies.

The most important of the publications issued by the Commerce Department is the *Census of Manufactures,* issued by the Bureau of the Cenus, which provides a vast amount of information on industries and products. Unfortunately, the *Census of Manufactures* is conducted only every 4 or 5 years. The last census covered 20 industry groups and 7 groups of chemicals and allied products. Information is reported on the number of plants, size of establishments, employment, raw materials cost, fuel and power consumption, inventories, production, shipments, and regional breakdowns.

In order to fill the gap between the publication of various issues of the *Census of Manufactures,* the Bureau of the Census issues an annual booklet entitled the *Annual Survey of Manufactures.* This survey is based upon samples and contains information similar to the *Census of Manufactures* for selected groups of primary interest.

To meet the need for promptly published current data on the output of individual products, the Bureau of the Census issues several series of *Current Industrial Reports* at monthly, quarterly, and annual intervals. The most important one is the series on inorganic chemicals, which gives monthly production data on over 50 chemicals, with breakdowns by grade and by process on the leading ones. Unfortunately, the type of detail supplied for synthetic organic chemicals by the ITC is not available for inorganic chemicals. Other series of interest to the chemical industry are the ones on superphosphates; paint, varnish and lacquer; gelatin; and animal glue.

Among the quarterly and monthly periodicals of bhe Business and Defense Service Administration (U.S. Department of Commerce), which present industry reviews and statistics, is the *Chemical and Rubber Industry Report.*

The monthly *Survey of Current Business* reproduces thousands of statistical series, with comparative figures for the 12 preceding months. The biennial supplement *Business Statistics* carries the series back for 5-25 years. Helpful data are also found in the annual *Statistical Abstract of the United States,* which reproduces condensed figures from government reports of every type.

The National Technical Information Service (NTIS) of the U.S. Department of Commerce is a central source for the public sale of government-sponsored research, development, and engineering reports and other analyses prepared by federal agencies, their contractors or grantees. It is also a central source for federally generated data files that can be machine-processed.

The NTIS information collection exceeds 900,000 titles; all are available for sale. The agency has an on-line computer search service that allows quick search for summaries of over 420,000 federally sponsored research reports completed from 1964 to the present time.

The NTIS Bibliographic Data File on magnetic tape, which includes published and unpublished abstracts, is available for lease, as are the computer products of other federal agencies.

Current summaries of new research reports and other specialized technical information in various categories of interest are published in weekly newsletters (*Weekly Government Abstracts*), which are indexed. An all-inclusive biweekly journal (*Government Reports Abstracts*) is published for those requiring all the summaries published in a single volume and accompanying index.

A standing order microfiche service is also available that automatically provides subscribers with the full texts of research reports selected to satisfy individual requirements.

Other services, such as the coordination, packaging, and marketing of unusual technical information may be specifically designed.

Federal agencies and others registering reports with the NTIS are: Department of Agriculture, Department of Commerce, Department of Defense, Department of Health, Education, and Welfare, Department of Housing and Urban Development, Department of the Interior, Department of Labor, Department of State, Department of Transportation, Department of the Treasury, Environmental Protection Agency, and the National Science Foundation.

It is not the intention of this section to describe each service in detail; however, an overview of the various services offered will give the reader a feeling for what is available. The weekly newsletters are filed under the following headings: Administration, Agriculture and Food, Behavior and Society, Biomedical Technology and Engineering, Building Technology, Business and Economics, Chemistry, Civil and Structural Engineering, Communication, Computers, Control and Information Theory, Electrotechnology, Energy, Environmental Pollution and Control, Government Inventions for Licensing, Health Planning, Industrial and Mechanical Engineering, Library and Information Sciences, Materials Sciences, Medicine and Biology, Natural Resources, NASA Earth Resources Survey Program, Ocean Technology and Engineering, Physics, Problem-Solving Technology for State and Local Governments, Transportation, Urban Technology.

The NTIS also offers a *Published Services* facility that offers computer services identical to the on-line custom searches also offered at a much higher price; these reports have been completed and published in response to previous demand and consequently are less costly for the NTIS to produce.

The Library of Medicine's MEDLINE/TOXLINE service is also available through NTIS. MEDLINE is an on-line data base maintained by the National Library of Medicine containing references to about half a million citations from 3000 biomedical journals and is updated monthly. TOXLINE is the National Library of Medicine's on-line collection of computerized toxicology information containing more than 375,000 references to published human and animal toxicity studies, effects of environmental chemicals and pollutants, adverse drug reactions, and analytical methodology.

An *Energy Standing Order Service* automatically brings the full text of every printed report in the field of energy. The reports are usually grouped under the following headings: Energy Sources, Energy Use, Supply and Demand, Power and Heat Generation, Energy Conversion and Storage, Energy Transmission, Fuel Conversion Processes, Policies, Regulations and Studies, Engines and Fuels.

A *Petroleum Statistics Report* combines the *Weekly Petroleum Statistics Report* and the *Weekly Petroleum Situation Report* into one comprehensive monthly periodical compiled by the Federal Energy Administration (FEA). The *Monthly Energy Review* is the principal communications medium for the FEA. The *Review* is divided into an overview, Energy Sources, Electric Utilities, Resource Development, and Price.

The *Selected Water Resources Abstracts Journal and Index* covers water pollution, water lay, ground water, lakes and estuaries, water yield, watershed protection waste treatment, water demand, hydraulics and soil mechanics.

Government-owned inventions are available to entrepreneurs, often through royalty-free, nonexclusive licenses. Some government agencies may grant limited exclusive licenses when it appears that such licenses are necessary for the attraction of risk capital. Royalties may be negotiated depending on the rights that are transferred. All government agencies submit their new inventions to NTIS at the same time that they file patent applications with the Patent and Trademark Office.

New government-owned inventions, patents, and patent applications are summarized weekly in NTIS's illustrated newsletter *Government Inventions for Licensing.* Inventions that are believed to have more than average commercial potential are highlighted in the newsletter as are those for which the government has foreign patent rights available.

Innovations resulting from National Aeronautics and Space Administration research are described in one or two pages. Items selected for review have potential for commercial application. Included are reports on new products and devices, manufacturing processes and techniques, and materials.

Naval Research Labs *Tech Progress Reports* cover work done by the U.S. Naval Research Laboratory. This publication summarizes technical progress on specific projects or problems on NRL's program as well as completed research studies. It reports on chemistry, electricity, metallurgy, nuclear, atomic and optical physics, ocean sciences and engineering, radar, solid state physics, and related fields.

U.S. exporters now have access to current multinational governmental regulations and codes of practice involving the sale of manufactured products worldwide. The Technical Help to Exporters (THE) service of the British Standards Institution provides this vital information through NTIS. Also included in this unique service are foreign approval and certification procedures and descriptions of the enforcing organizations throughout the world. The information is gathered worldwide on-the-spot by THE specialist engineers and reported in technical digests, special investigation reports, information sheets, translations, and annual directories. Use the order form on page 73 to request the Technical Help to Exporters Catalog (NTIS-PR-244) for more detailed information.

Half of the world's scientific literature is published in non-English languages, two-thirds of which is in Russian, Chinese, and Japanese. Acting under congressional authorization the National Science Foundation coordinates and administers the Special Foreign Currency Science Information Program (SFCSIP) for the federal government, which makes selections for special translations.

Besides the above sources of information, other sources can be helpful on certain assignments. The Library of Congress maintains extensive research facilities that can be of value. Although most often the information desired is on production and plant capacities, many times data are required on consumption trends and forecasts. Here the economic studies of the Department of

Agriculture can be very useful. Also, such agencies as the U.S. Bureau of Mines and the Alcohol Tax Unit of the Treasury Department publish helpful data on consumption of various products.

It is often possible to obtain information for one purpose that was originally published for another. For example, plant capacities can sometimes be estimated from applications for reductions in freight rates for commodities. The Securities Exchange Commission requires corporations to make public a wealth of information on their operations in connection with the sale of new securities. In addition, the various war and defense agencies frequently release basic information about individual industries and trades in connection with their various production-control orders and allocations of raw materials.

Much information is available from the published hearings before such groups as the Food and Drug Administration and the Federal Power Commission. Hearings conducted by various congressional committees, together with their reports and recommendations to Congress, also provide valuable data. This material often requires patient digging, but personal contact with people in the affected agencies may quickly uncover the pertinent data. Detailed information is also available from the annual hearings before the congressional appropriations committees. Various research and development programs are sometimes summarized in the annual reports of the chiefs of the various bureaus.

Literature Sources

Literature sources, although not considered to be useful by many commercial development personnel, are anything but that. Much valuable information can be obtained that is of great assistance in the product development effort.

Literature sources can be divided into the following types:

- Trade journals
- House organs and product bulletins
- Annual and 10K reports, prospectuses and proxy statements
- Financial institutions
- Court records and congressional reports
- Indexes and bibliographies
- Directories
- Government publications
- Trade association publications
- Regional reports

Government and trade association publications have already been discussed; the other sources are reviewed in the following sections.

Trade Journals

Constant reading of trade journals is highly recommended. Naturally it is impossible to do more than scan the available literature in certain fields of interest, but with reliance on abstract services, it should be possible to limit one's reading to a few selected periodicals. A number of these journals carry so-called depth studies. These studies should not, however, be accepted at face value. Without derogating the ability of the personnel associated with these journals, it should be recognized that they cannot possibly do the same type of depth analysis as can an in-house group, because they lack the time to make the large number of contacts necessary to perform the kind of work that will satisfy the demands of senior management. At best, these articles are useful guides to further work, but when done badly, they are misleading and dangerous. They do, however, perform the useful function of focusing attention on areas of potential interest.

House Organs and Product Bulletins

This source is especially useful for learning about new uses for products offered by a company. In addition, information about a company's attitudes toward marketing and other information such as its research and development organization, history, and areas of general interest are often made available. The researcher should, of course, differentiate between the so-called "puff" and factual information.

Product bulletins are also a valuable source of information. Companies often prepare comprehensive brochures discussing actual and possible end uses for their products, as well as chemical and physical properties, reactions, patent information, and specifications. The researcher can, of course, be led astray at times, because companies often claim end-use applications for their products that either have become obsolete or are as yet only in the laboratory stage. Nevertheless, this source can be helpful in filling gaps in the researcher's knowledge.

Annual Reports and 10K Reports, Prospectuses, and Proxy Statements

Annual Reports and 10K Reports

Because most of the larger chemical companies are publicly owned, they must file annual reports, 10K reports, and, in most cases, abbreviated quarterly statements.

These reports vary considerably in comprehensiveness and detail, but all are required to reveal such information as sales, profits, dividends, taxes, balance sheets, and income statements. Also included in many reports are lists of plant locations and their products, new products, discontinued products, expectations for the future, discussions of the company's performance, its problems and opportunities, and lists of officers and directors.

Prospectuses

Prospectuses constitute an excellent source of company information and are much more detailed than company annual reports, because the Securities and Exchange Commission requires complete disclosure pertinent to new issues. They furnish considerable detail on a company's product line, historical background, breakdown of sales by product groups for selected years and distribution channels, and they often give information on a company's estimates of its share of selected markets. Also presented are data on current research projects, patents held, and considerable financial information.

Proxy Statements

Proxy statements sent to stockholders are available from the issuing company as well as from libraries. These statements must provide the names of the directors and the money paid to them, the three highest paid officers whose direct remuneration exceeded $30,000 in the previous year, a list of nominees for the board of directors showing the amount of stock held by each, and the direct remuneration paid to all directors and officers during the last fiscal year.

Financial Institutions

Many banks publish valuable newsletters, usually on a monthly basis. These are most often concerned with general business conditions, but many offer data of use to the researcher. Some examples are:

Business in Brief, Chase Manhattan Bank
The Petroleum Situation, Chase Manhattan Bank
Report on Western Europe, Chase Manhattan Bank
Monthly Letter, First National City Bank of New York
Business Review, Federal Reserve Bank of Philadelphia
Business Conditions, Federal Reserve Bank of Chicago
Monthly Review, Federal Reserve Bank of New York

Court Records and Congressional Reports

There have been several instances in which court cases or congressional committee reports have provided a bonanza. Data of possible interest are read into the court record by virtue of two types of actions: infringement suits and antitrust actions.

Patent infringement suits generally involve one product or product type. Two or more companies may be involved in such actions, and the information brought forward may be for one or more individual companies or for an entire industry, depending upon the nature of the suit and the structure of the industry. Infringement suits occur most often in product areas of great technological flux.

Antitrust suits generally yield data of an industry-wide nature, often in extreme detail. The same is true of the occasional investigations into the affairs of one industry or another.

Copies of reports issued by congressional committees and by the Federal Trade Commission and similar bodies can often be obtained by the public. The same is not true of court transcripts. Here it may be necessary to visit the court in which the case was heard in order to read the transcript.

Indexes and Bibliographies

Indexes and bibliographies can be of great assistance in keeping track of the multitude of journals published each year. A number of catalogs are available, of which the most prominent are:

Business Periodicals Index

Public Affairs Information Service Bulletin

Applied Science and Technology Index

The Engineering Index

The New York Times Index

The Wall Street Journal Index

Monthly Catalog of United States Government Publications

Directories

There are many directories that cover a wide range of subjects. Some of the major directories are the following:

Guide to American Directories

Federal Statistical Directory

Guide to U.S. Government Serials and Periodicals–Documents Index

Standard Rate and Data Service, Inc.

Regional Reports

Industrial directories are issued for many states and cities. These classify firms by industries, alphabetically and geographically. Statistical bulletins are published by most state departments of agriculture; these report the size of the principal crops and show fertilizer consumption in detail, by county and product. State bureaus of mines issue bulletins with much valuable text and statistical material on the production of minerals. Most states have departments of commerce, and most state universities have bureaus of business research that publish valuable reports and assist in locating information.

Chambers of commerce are also a prime source of regional data. For example, the Baltimore Association of Commerce, the Dallas Chamber of Commerce, and the Houston Chamber of Commerce publish monthly graphs and statistics of business trends, economic reports, and articles on local affairs and industries.

Banks also offer a great deal of regional information. Each of the Federal Reserve banks gives reports on general business conditions and current economic developments. Large private banks also perform much the same service.

Bibliography

Bateman, Robert I. "External Sources of Commercial Chemical Development." Speech delivered at a symposium on Commercial Chemical Development, New York, 1970.

Blankenship, A. B., and Boyle, J. B. *Marketing Research Management.* New York: American Management Association, 1965.

The Conference Board. *Consultants: Selection, Use and Appraisal.* New York: The Conference Board, 1970.

The Conference Board. "If I Were The Client . . ." *Experiences in Marketing Management,* no. 21, 1969. National Industrial Conference Board, Inc., 845 Third Avenue, New York, N.Y. 10022.

——. *Using Marketing Consultants and Research Agencies.* New York: The Conference Board, 1966.

Corley, Hoyt M., ed. *Successful Commercial Chemical Development.* New York: John Wiley, 1954.

Federal Statistical Directory. Superintendent of Documents. U.S. Government Printing Office, Washington, D.C. 20402.

Frankenhuis, Jean Pierre. "How to Get a Good Consultant." *Harvard Business Review,* November-December 1977, pp. 133-139.

Giragosian, N. H., ed. *Chemical Marketing Research.* New York: Reinhold, 1968.

——, ed. *Fundamentals of Commercial Development.* Symposia presented under the auspices of the Chemical Marketing and Economics Division, American Chemical Society, New York, 1969.

"Just Among Us Consultants." Conference Board Record, May 1973, pp. 61-64.

Shay, Philip W. *How to Get the Best Results from Management Consultants.* New York: Association of Consulting Management Engineers.

United States Government Manual, published yearly. Superintendent of Documents, U.S. Government Printing Office, Washington, D.C. 20402.

The Use and Misuse of Business Consultants. Symposium presented under the auspices of the Chemical Marketing and Economics Division, American Chemical Society, Detroit, 1965.

PART III

INTERRELATIONSHIPS

12
WHAT TOP MANAGEMENT EXPECTS FROM COMMERCIAL DEVELOPMENT

James G. Affleck
American Cyanamid Company
Wayne, New Jersey

The role of commercial development has changed substantially over the past 15 years from a product-oriented function or process to one involving the development not only of new products but also of new business.

During this period considerable confusion has arisen concerning the role a commercial development department should play in the organization. The nature of many commercial development departments has changed with the advent of planning departments and the expansion of marketing research functions.

As stated in the introduction, commercial development cannot be carried out effectively without the complete support of the chief executive officer who is, in effect, the head of the commercial development effort.

This chapter will discuss what top management expects from commercial development in the light of the changing industry.

In a study conducted by Heidrick and Struggles, 49 company presidents were interviewed to determine their attitudes toward commercial development. The interest on the part of chemical company presidents is shown by the fact that 44 of their 69 questionnaires were completed and returned. Heidrick and Struggles guaranteed anonymity for anyone who wished it by using unmarked pages and self-addressed, stamped reply envelopes. Several questionnaires were returned unsigned, but most were signed, and several wrote letters of up to two full pages in length. The questions asked and the responses were as follows:[1]

In the comments section, the presidents stressed a number of points needed to increase the effectiveness of commercial development in the chemical industry:

1. Necessity of greater profit orientation in evaluating commercial development projects.
2. Greater consideration of cost and availability of capital funds.
3. Presentation to top management of more and better-documented recommendations.
4. An improvement of communications between top management and the commercial development function.

The questionnaire consisted of three multiple-choice questions and two discussion questions. In framing these questions Heidrick and Struggles had to choose between something short enough to get the attention and cooperation of busy people and digging deeply enough into their psyches to make a profound contribution to commercial development. The short route was chosen.

Question 1 asked: "Over the past five years, how has the contribution of your company's commercial development activity compared with your expectations?" The three choices which were offered as answers to these questions were (1) exceeded my expectations, (2) met them, (3) failed to meet them. In answering this question, 3 said their commercial development department had exceeded their expectations, 16 said they had met them, and 25 said they had failed to meet them. Later on, I will elaborate on this.

Question 2 asked: "Over the next five years, how would you rate the importance of commercial development to your company, compared with the importance of the function during the last five years?" The three choices offered were (1) more important, (2) about the same, (3) less important. Of this group, 28 said commercial development would be more important over the next five years, 10 said it would have about the same importance, and 4 predicted less importance.

Question 3 asked: "Is commercial development currently offering a good opportunity for management development in your company today?" Of the group, 33 said "yes," and 11 said "no." One president said: "Commercial development should be a good management development area, but we can't get our people to move out of there." He admits he might be mistaken.

The fourth question was deliberately an open-ended question: "What should your commercial development department be doing to make it more valuable to your company?" One of the presidents who said commercial development had exceeded his expectations made the point that his commercial development department had redirected some of its efforts toward the purchasing and decision-making groups in customer companies rather than spending all its time with research and technical people. The other two presidents seemed to feel that their commercial development departments were doing well enough and that no improvements were needed.

Among the people who said that commercial development had met their expectations over the past 5 years, typical comments to question 4 were:

> "That elusive index called market potential needs to be defined better for each project. Generally, commercial development people need to develop a keener appreciation of existing and probable competition."
>
> "New business development is vital to our company's survival. We must, therefore, define corporate goals for 5 and 10 years in the future, evaluate our present and obtainable abilities, technology and strengths, and then concentrate on capitalizing on the possible."
>
> "A more *practical* approach to enlarging the scope of present products and closer touch with research and development on improvements."
>
> "Develop communication with all our company management to encourage the free flow of ideas."
>
> "We need a more sophisticated project evaluation system. Commercial development needs more initiative."
>
> "More careful screening of ideas to minimize cost."

Throughout the replies to these questions there was a distinct plea for "reality" on the part of commercial development people and a plea for the assumption of more responsibility to *complete* projects and come to management with well-documented recommendations either to do something or to do nothing. There were several comments that could be condensed to the simple request for commercial development people to be more creative. In other words, it is not only appropriate but extremely profitable to concentrate on trying to develop fresh ideas no matter how difficult that may be. With the changes in our economy resulting from fuel shortages, ecology pressures, and the constantly changing value of international currencies, there is an excellent opportunity for the person who can critically reappraise his or her company's capabilities in terms of today's economic and social conditions. The next 5 years appear to be years of change in terms of both raw materials and markets. The last item read: "Please add any comments you care to make on the subject of commercial development." Some of the comments were:

> "A very important function when it makes profits come to pass; less important when it concerns itself with projects without regard to profit."
>
> "Commercial development must become international in scope in the future."
>
> "Commercial development should keep the cost and availability of capital funds in mind when starting new projects because capital is going to be a major problem throughout the next 5 years."

"Speed in commercial development is vital in establishing success or failure. Without speed, projects which are ultimately successful take too long to implement and projects that are ultimately failures end up being more expensive failures."

Some presidents were negative about their commercial development departments. One said, "I think it's an obsolete term." Others said they feel that commercial development people live in a world of their own and forget that they are part of the "home team." To some extent commercial development people are subject to the criticism often leveled at basic research people. Some researchers are accused of being interested in research but not in whether or not their company makes a profit. The president who was the most critical of his commercial development department admitted that it could be partly his own fault because he was sure the department had never had specific objectives outlined for it.

My View of Commercial Development

Corporations, like the people who manage them, change. The corporate personality is altered by growth, by new directions dictated by markets and products, and by alteration of the environment in which it exists. To survive, the corporation must be capable of responding to new *environmental* pressures by its ability to reshape, expand, consolidate or change direction entirely. Stimulation comes from customer demand, fluctuating economic factors, governmental pressures, and the attitudes and philosophy of current company managers.

Change occurs more slowly in the corporation's operating divisions because modification of technology and products takes place over the long term. Change is more frequent in staff functions because these are primarily people changes. Furthermore, staff functions influence, and are influenced by, company philosophies and attitudes and thus more closely reflect current management thinking. Nowhere is this more evident than in corporate planning. An indication of this change and expectation of even further modification is the recent renaming of our former Commercial Development Division. It is now more accurately called the Corporate Development and Planning Division.

Cyanamid management expects planning and commercial development to set for itself an entirely different set of objectives than it did 10 or 15 years ago.

To some degree the corporate planning group is expected to be the company "think tank," or its intellectual center, although we hope the think-tank concept or an intellectual approach to business decisions is not limited to a single staff function. These are attitudes that we expect to see in every division and segment of our business. But through the nature of its functions and the talents of the people who staff it, we do expect the planning and commercial development group to set the pace and to act as an intellectual catalyst in the planning process in long-range decisions among all of our segments. In this, the planning group becomes the pioneer and the leader for other units to emulate.

Only a few years ago, commercial development was charged with responsibility to operate across the board in determining how the company could commercialize its capabilities in manufacturing, marketing, and research. It was organized to function across product and geographical lines in such areas as research management, market research, market development, new business and financial planning, almost all of it based on long-range market forecasting.

Although these basic approaches are still in force, we have changed our thinking somewhat. New business ventures, new technology and products, and new market entries still form the backbone of corporate growth. But we have concluded that we can no longer make these decisions in a "hothouse environment" and expect them to be transplanted successfully to the cold, competitive world of our operating divisions.

The reasons are not very difficult to understand. Most successful new business ventures are really those that grow as extensions or modifications of already established businesses. Experience has shown us that decisions in these areas can be more accurately made if the new ventures are thought of as children of our operating divisions, which already have their own research, planning, and marketing experts who are specialists in their own particular fields.

As a result, we have transferred much of the responsibility for commercial development from our corporate staff to the operating divisions.

This decision was forced primarily by that very powerful influence of every human institution called human nature. The medical products manager, the agricultural marketer, the chemicals researcher all believe they know far more about their products, technology, and markets than anyone in a staff division responsible for commercial development. Who can argue with them, because they enjoy the vantage point of day-to-day contact with a single product and single market area while the staffer must be a generalist, dividing his or her time among all of the parent company's spheres of interest? And now much of the former commercial development responsibility is in the hands of the experts who face the difficult task of proving the wisdom of corporate planning decisions in the marketplace.

This has freed the Corporate Development and Planning Division to devote its energies to another critical function—to guide and evaluate divisional planning efforts.

Despite the ability of each operating division to define a pattern of growth within its own specialized area, there is still a vital role for the corporate planning group, because the strategic development of a corporation should not simply be the aggregate of the development plans of all of its divisions.

Cyanamid, as an example, now looks to individual divisions to determine where our opportunities lie in relation to our present businesses.

But we often find that by adding these divisional growth plans together, they seldom provide the growth potential we have set for the entire corporation. We find overeagerness from some and too much conservatism from others, or we discover that we are not, as a single entity, likely to reach our primary objectives.

When you sum up divisional objectives and step back and view them from an overall managerial vantage point, you almost always find that some company resources are to be strained severely and others underutilized.

To supplement divisional planning, therefore, we have assigned specific responsibilities to our Corporate Development and Planning Division.

Let's define these functions before we discuss what we can expect in the future for the development and planning group. They are:

1. To provide guidance to operating divisions. The Corporate Development Department defines the nature of development opportunities that are needed to meet overall corporate objectives. The corporate group is in a better position than any single operating division to analyze where we have the most abundant resources, including capital, and where our resources are deficient. It is also in the best position to predict what new resources will be needed and available in the future.
2. To strike a balance among divisional proposals. The Corporate Planning Department is expected to propose a mix of available options to achieve corporate targets and to recommend additional elements needed to strengthen that mix.
3. To advise senior officers about what ought to be their expectations for the future, based on the available and anticipated resources of the company. This assessment tells senior managers what they should expect in terms of investment rates, financial policy, capital needs, sales, earnings, personnel requirements, and other factors.
4. To advise senior officers and the operating divisions of how technological, social, and political changes will affect the company, and to advise on what actions should be taken to capitalize on opportunities or sidestep any threats that could result from these changes.
5. To identify areas where we can do something to create changes that work in our favor. This involves an active rather than passive role in the business environment in creating new markets and new technology to correct potential problems of society—and the market—before they occur.
6. To identify completely new business entries, while at the same time maintaining the ability and initiative to identify business exits. Leaving a business at the right time can sometimes be far more profitable than adding to diversification.
7. Be a window on this very competitive world. Corporate development and planning must be able to provide accurate, detailed reviews of who our competitors are. A first-class operation will also be up to date on the

apparent strategies, objectives, resources, and limitations of our competitors. Planners should be kept advised on how competitive factors are likely to impact upon our own strategies and opportunities.

8. Serve as a source of skilled people who are capable of moving up the management ladder in an operating division. Development and planning should serve as an attractive recruiting tool. The nature of its work is appealing to bright, highly motivated people and provides a day-to-day working environment that offers excellent training, both in terms of the specifics of business, financial, and marketing decisions and in seeing firsthand how decisions are made at the senior level. It is an effective starting point for young people on the way up.

9. Finally, to provide an information resource. Corporate development and planning is an ideal place to collect and organize commercial information and other data useful in planning decisions. It is an ideal repository for competitive and technological surveillance to be made available to all segments of the company.

If this then is the present, where is the corporate planning function headed in the future?

Here I must repeat something I said in another recent analysis of the kind of company Cyanamid and others like it will be in years to come.

The future of management may well be "nonmanagement."

More than ever, the route to success will be the development and utilization of people organized to employ *all* of their individual creative talents to the maximum, within an environment of continuous and dynamic change.

Priority will go to the motivation of people in order to extract the utmost from products and technology. The rigid and highly structured organizational framework of the past will be replaced by a cohesive interdependence of thought and action, perhaps without conscious direction or apparent leadership as we now understand it. Management's main job will be to exercise sensitivity and an educated intuition to draw the maximum from a highly skilled and intellectually sophisticated force of managers and workers who will probably possess more specialized skills than ever before.

It is axiomatic that any organization, whether business or government, exists to perform a specific function, or it should not exist at all. We often fail to realize, however, that the organization's function cannot be static under the pressures of business competition.

With growing regularity, business enterprises must be capable of abrupt starts, stops, and turns. Pressure from competitors, government, consumerists, and others often force business abruptly into new areas for better or worse. Research and marketing developments prompt major changes. No organization chart can be immutable; rather, the organization must be flexible enough to adapt continuously to the external environment to remain productive and useful to society. The alternative is to cease to exist.

This modern dynamism has major implications for the structure of organizations and how they should develop the human and nonhuman resources that are their basic assets. In particular, it will place greater stress upon the corporate development and planning function.

Our past methods of reaching short-and long-term objectives through methodical, step-by-step progress, usually with various specialties and functions separated until drawn together by the master puppeteer labeled "chief executive" will be replaced by a new, much less structured way of life.

In such an environment we can expect all aspects of management's job to be more subjective and complex, with more shades of grey to analyze in solving any problem. It will be, in many ways, the constructive orchestration of chaos. To those of us trained in the old ways it will seem by comparison both inconsistent and paradoxical, without the logic of neat pyramids of organizational boxes and the territorial imperatives that have grown up around and control so many corporate functions.

As part of this evolution, corporate development and planning, because of its unique nature, will continue to be a very important function. This area of expertise is very unusual because it involves much less specialization than such other staff functions as research, manufacturing, and marketing. But it is because of this lack of specialization and the ability to adapt to the general, ever-changing needs of the corporation that its role will continue to increase in importance.

The scope of its responsibility and the nature of the experts who staff it make development and planning eminently capable of accepting the integrative, anticipatory role that it must play in the fluid, dynamic type of corporation that we foresee.

While management at all levels in the operating and staff divisions will probably become more freewheeling and less restricted by current modes of decision making, there is every reason to anticipate that there will be increased specialization within these divisions. Such greater specialization within separate company units can be expected from the increased complexity of products and of the global market we serve.

As a result of these complexities, the role of the part-time expert will diminish. We have already modified the role of corporate development and planning in divisional long-range planning. So we can expect that senior executives and operating divisions will look with decreasing frequency to corporate development and planning for advice, insight, or specific services related to their own business areas.

At the same time, the corporate planner's role as a functional expert in strategy development and the formulation of policy will continue, but with some modification.

Because corporate planners cannot carry out the function for which they offer advice and guidance, they will have to assume the role of educators within the corporation. They will have to accept more of a discoverer's role. They will be the ground-breakers in a learning institution where the student body stays the same, and the curriculum grows.

In short, members of the Corporate Development and Planning Division will become the company's "futurists." They will make study, understanding, and control over the future a matter of science and not of art.

In the past and now, commercial development has borne the task of questioning and evaluating what is going on and what is being proposed. Its stock questions are: "Is it prudent?" "Is it wise?" "Is it well-balanced?"

But in the future, corporate development and planning questioning will occur at an earlier stage, dealing not with the world that is, but with the world of what might be. Corporate planning will prod or guide operating managers and researchers in directions responsive to changing stimuli.

Planners will also have to shift increasingly to the role of interpreter not only of trends, data, and proposals but also of a world increasingly populated with specialists at functioning levels of increasing minutiae. The role of interpretation will be made more critical because of the narrowing viewpoint of the managers it is intended to serve.

As futurists, corporate planners will be charged with seeing that the corporation extracts maximum productivity and optimum use of its human, financial, and physical resources. It is a role essential to the survival of the corporation and to the success of its primary role: to supply society with the products it needs and wants at a price it is willing and able to pay.

Note

1. These results were presented in a speech given by John S. Wilson, Vice-President of Heidrick and Struggles.

Bibliography

Alcorn, David S. "How Management Looks at Commercial Development Today." Talk given at symposium entitled, "Evaluating the Effectiveness of Commercial Development," June 1971.

Barrett, William R. "A President Looks at Commercial Development." Talk given at symposium entitled "Evaluating the Effectiveness of Commercial Development," June 1971.

Caldwell, C. G. "Priorities in Commercial Development Today." Talk given at symposium entitled "Evaluating the Effectiveness of Commercial Development," June 1971.

Canavan, William J. "What Management Expects from Commercial Development." Talk given at symposium entitled "Changing Times in Commercial Development," March 9, 1971.

Donley, Edward. "Commercial Development at Air Products and Chemicals, Inc." Talk given at symposium entitled "Changing Times in Commercial Development," March 9, 1971.

Fleming, Samuel C. "Trends in Commercial Development for the 70's." Talk given at symposium entitled "Changing Times in Commercial Development," March 9, 1971.

Gerlach, John T. "Fashion in Commercial Development." Talk given at symposium entitled "Evaluation of New Projects and Successful Techniques to Implement Them," November 1, 1972.

Werner, Jesse. "Scope of Commercial Chemical Development–Past, Present, and Future." Talk given at 23rd Annual Meeting of the Commercial Chemical Development Association, March 14, 1967.

Wilkinson, R. F. "Planning for Commercial Development Today." Talk given at symposium entitled "Evaluating the Effectiveness of Commercial Development," June 1971.

13
INTERRELATIONSHIPS WITH OTHER CORPORATE FUNCTIONS

David S. Alcorn*
Dyes and Chemicals Division
Crompton & Knowles Corporation
Fair Lawn, New Jersey

With the wide differences existing in company organizations today, commercial development may be located in various parts of the organization. In only a few companies does commercial development report directly to the top operating executive as was recommended in the commercial chemical development textbook published some years ago by the Commercial Chemical Development Association.

This discussion does not deal with where Commercial Development is located in a company or how it reports and is managed. It deals with those relations among people carrying out specific functional duties in order to get new products successfully introduced to the market in a profit-making organization no matter how it is organized.

All discussions of new product development emphasize the high risk involved and the need for teamwork across all functions of the business. Good team effort ensures that the new product performs a needed function, is properly priced, is properly presented to prospective users, is available in quantity to meet demand, is properly packaged, and so forth.

Equally important, a properly functioning team can spot failures at an early date to save money and to get more successful projects going. We shall

*Affiliation at time of writing: Executive Vice-President, J. T. Baker Chemical Co., Phillipsburg, New Jersey.

try to look at the working relations that are essential to creating an effective team effort and success in new-product activities.

Interrelationships with Corporate Management

The most important relationship of commercial development is with corporate management or the top operating management of the company. If corporate management has written statements defining corporate policy and goals in some detail, this is a great help to development groups. Unfortunately in many cases corporate policy in regard to new products or businesses is not sufficiently articulated, and this is a cause of many problems.

New projects almost always involve a heavy drain of money, particularly if new plant investments are needed. This drain of money goes well beyond the normal 1-year budget requirement. However, when the project starts, a commitment, perhaps to be made 3 years later, is implied at the very beginning. Frequently, it appears that the commercial development group does not properly stress the need for future investment if the project is successful, or this need is not "heard" by appropriate management, perhaps because the money is not a part of the current budget. The lack of understanding is only recognized after some period of time has passed and considerable effort and money have been expended. It is of prime importance that the full commitment and understanding of management be obtained at the start of the project, and this understanding and commitment needs to be reviewed at frequent intervals during the life of the project.

There are many reasons why a seemingly attractive project does not enjoy continued support of corporate management. Some key questions that should always be asked to explore potential areas of disagreement are the following:

1. Will we make the necessary investment to carry the project to a large scale if the first efforts are successful?
2. Does staffing the project (labor and supervision) represent any problem in terms of relationship of number of people to sales dollar or type of skills needed? If the project does not match normal corporate measurements, a full explanation should be made.
3. What performance measurements will establish credibility for the effort and build sufficient rapport to ensure continued support in difficult periods ahead?

A lack of understanding of management concerning what is involved in the project is probably the major factor in poorly spent money, ineffective results, and poor morale among development groups. Frequently one hears development groups speak of projects that were going well but were unexpectedly killed. It has been my experience that much of the problem comes because management does not spend enough time on the project to understand it until they are faced with a need to make a major investment. At that time, management may decide they were never interested in the project in the first place, with disastrous results for all those involved.

Communication with management can be improved if essentially all reports are placed in a standard format, one that is similar to the format used for managing the existing business. This makes it easy for management to make comparisons between projects because they are all looked at on an equivalent basis and also makes it easier to compare a new business activity with the existing business. There are two major reports that every commercial development organization must prepare. The first is the new product or new project report, which is the first report to management proposing a new venture; it generally asks management's approval and financial support. The second report is a status report that on a regular basis reports progress in the projects. The outline below lists a standard format for a new product report that has been used with some success at J. T. Baker Chemical Company.

New Product Plan

1. *Background*
    a. Nature of the market–brief description of the market and use of the product
    b. Market size for the product (generally presented for 5 years out with indication of projected pricing)
    c. Product features needed and competitive situation
    d. Proposed process
    e. Buyers
2. *Objectives*
    a. Market position
    b. Sales and profit goals–generally presented as projection for 5 years out
3. *Strategies*
    a. Approach to securing business
    b. Pricing
4. *Estimated project costs*

    Research costs–projected through trial plant run

    Capital expenditures

    Manufacturing expense
5. *Action plans*

    Includes timing to get product on the market

A second area where understanding of management is essential if commercial development is to be successful over a period of time involves the proper staffing of projects and the proper treatment of the people working in them.

Any new project needs dedicated people who will spend some significant portion of their career to ensure the success of the project. If management moves out promising performers who may be needed to solve a current problem it has two bad effects:

1. It hurts the project because the performance of that individual is missed.
2. It communicates to everyone involved in the project that management is more concerned in meeting today's problems than in ensuring the future of the business.

Management must realize a new project that the individuals who are entrusted with the project must be highly skilled in the field involved. Generally, this is taken for granted in obtaining research and development personnel because their entire career is involved in new projects. The need is not always recognized in regard to marketing personnel or financial or production personnel who can contribute to existing business. Top-flight performers in these areas are needed in new projects, and good performance must be adequately rewarded regardless of success or failure of these high risk projects.

Interrelationships with Research and Development

The problems of relationships between commercial development and the research and development function have been dealt with at length at meetings and in the literature of commercial development. Frequently commercial development is under the direction of management of research and development, and there are certain advantages in the relationship. The following guidelines are suggested where the two functions are separate:

1. Research and development must be given a clear understanding of the basis for the company's interest in the new project. Is it to augment existing product lines, to capitalize on existing capacity that is not utilized, or to seize some new market opportunity? It is essential that both the commercial development group and the research and development group have the same goal in mind, and this problem of common agreement on a goal seems to be most difficult between commercial and research people. You all remember the definition of a fanatic—an individual who has redoubled his efforts after forgetting what the goal was. None of us need efforts of this type.
2. A clear understanding is needed of the economics and profit goals of the project. The research program must be committed to a goal of obtaining

a product that shows a profit. We all hear too frequently of projects that were technological successes but commercial failures.

3. The success of many projects depends on a solution to some technological problem or problems. In these cases the commercial development group should prepare as best they can a well-written definition of solutions that will lead to business success. Sometimes a research group has a technical solution that for one reason or another is overlooked because of too narrow an understanding of the goal. For example, we may have said that to be economically competitive, we need a process with 95 percent yield. However, at a later stage, we recognize that a 90 percent yield with a high throughput and slightly different product mix will make us competitive. Factors like this are difficult to anticipate. However, it must be remembered that many outstanding processes are based on a shrewd economic understanding of what it takes to make money, and these processes have been overlooked by individuals who did not properly understand all of the ramifications.

4. The research organization should accept the responsibility to develop supporting-use information as well as the product and process needed. This is taken for granted in pharmaceutical companies where a statement of intended usage and dosage is always presented with the product. In many industrial operations the research department seems to be willing to develop the applications but has a mental block when it tries to write procedures for the use of the product because this is sometimes viewed as being a marketing or advertising function. But without research participation this information is apt to be superficial at best.

5. Frequent written communications between commercial development and the laboratory group spelling out activity and progress made are essential.

6. A climate must be created to induce acceptance of useful technology whether or not it was developed within the company's laboratory. In other words, we must fight constantly the "not invented here" syndrome. No matter how much money a company spends, it is a small part of the total research and development work done in the field in the world. The goal should be to make the best possible use of any useful technology.

7. During the initial customer trials of a new product it is essential that the research department be closely involved and if possible to be present during initial customer evaluation.

8. There should be a written policy covering the company attitude toward publishing research results. Where there is no such policy, disagreements between departments and much lost effort frequently result. The research department publishes papers that may seem to the commercial development group to be quite unrelated to supporting the development of new business. Sometimes valuable research papers are not published because of a fear that such publication will divulge company secrets, when in reality the publication would support the sale of company products. A clearly spelled out policy solves these problems.

Interrelationships with Production

Communications between the commercial development group and the production organization become more important at later stages of a project and frequently do not get sufficient attention. There is a real conflict between the normal production goals of low-cost, efficient production and the innovative process. This conflict can lead to poor results when attempts are made to produce a small initial volume of a product in existing facilities. For the commercial development person, these first production materials may be essential to the success of the operation. However, production of these materials may be a nuisance to a volume-oriented production operation. If the production individuals are told of and understand the future opportunity for the Company, their attitude toward difficult, small-volume production can be changed.

One way to do this is to communicate to production future estimates of sales and profitability with the request for the first small-volume production quantities.

It is of great value if the production department is in contact with the laboratory group while the process to make the new product is being developed. Much good information can be developed in the laboratory that can simplify the problems of initial plant production and reduce the extremely high costs that are frequently encountered.

With today's concern over pollution and environmental problems it is essential that this aspect of a new process be considered early in the game. This is particularly the case if research hopes that the new process will be fitted into existing equipment to avoid an initial capital expenditure. What may appear in the laboratory to be a very minor degree of pollution may be totally unacceptable if added to a plant situation where treatment facilities are being pushed to near and absolute limits. New federal and state regulations on materials that were not viewed as serious contaminants in the recent past can completely change process economics. For example, new regulations for ammonia content of waste water are now bringing the allowable levels down to parts per million. This in itself could eliminate an otherwise advisable approach toward certain organic compounds.

Interrelationships with Marketing

The fate of a new product in the hands of the marketing department depends a great deal on the climate within and outside the company at a particular time. Marketing people are more apt to be interested in a new product and sympathetic to the efforts of commercial development if the company is doing well. If performance has not been good, the new product effort is resisted as detracting from needed effort on existing products.

New product efforts may result in some competitive problems with existing customers. Good communication between commercial development and marketing results in better understanding and handling of such situations.

A problem can exist in staffing properly for commercial development efforts as was mentioned earlier. Unless definite efforts are made to keep good people in market development, they will in time become involved in support of existing businesses.

If a new product is to be sold through an existing sales force in the company, communications with sales mangement become of extreme importance. It is essential that the commercial development people understand exactly how the sales staff is compensated and ensure that the new product will fit into the sales staff's work load so that it represents a chance for them to increase their income in line with their opportunities selling other company products. Sometimes the intensive effort needed to sell a new product is not recognized, or an attempt is made to get extra sales effort without paying for it. This usually backfires in that good sales people are very much aware that time is their principal commodity, and excessive time spent on any one product must yield commensurate results. If a sales force is on a salary basis, recognition of sales successes with new products can be a powerful motivating factor, but this recognition must come from the line sales management that is responsible for the advancement of the sales staff rather than from the commercial development effort, which may be viewed as a staff group. If the sales force works on a commission basis, the commissions paid on the new product must cover the time required to establish it in the market.

Interrelationships with Purchasing and Finance

Frequently, commercial development people do not take full advantage of the purchasing function of their company. The purchasing department can supply more than the prices and availability of raw materials that are needed for the project. They frequently get much valuable information concerning market conditions, and, if they are kept informed as to the project goals, they can be very helpful indeed. The time spent in keeping purchasing informed of the status of new projects can be extremely valuable in terms of their success.

The involvement of purchasing people in make/buy decisions for new products requires great skill. Before undertaking a make/buy study for a product, the company market and profit goals should be very carefully spelled out and the efforts of the purchasing department should be guided toward reaching company goals rather than purchasing material at the lowest possible price. There are occasions when an aggressive make/buy study keeps a company from producing a product or entering a market in which it really should participate. The low-priced supplier may be seeking marketing position as much as he is seeking an opportunity to profit from the sale of the particular product. Sometimes an aggressive make/buy purchasing activity brings new producers into the field and effectively kills any opportunity for the company to make a profit from it. In the lack of well-spelled-out company goals this can be viewed as "good purchasing activity" despite the obvious adverse effect on the company's growth and profit position. Conversely, a make/buy study can keep a company from making some very unfortunate investments.

Closer involvement of financial management in new-product activity offers greater opportunity for better management of these projects in the chemical industry. For many reasons individuals with engineering background feel that they do not need, and indeed do not welcome, the consultation and advice of financial management whose training is primarily in accounting and related techniques. However, the mere fact that good financial people ask questions from a somewhat different point of view can ensure that the project stays close to reality. Financial management can be most helpful in dealing with problems of the following type:

Impact of investment now versus higher costs later

When to make the big investment

Detection of costs that are starting to get out of control, which will ultimately result in heavy overruns

Development of realistic research budgets

Focus on a profit goal rather than the technical achievement.

Financial people are frequently left out of new project meetings unless a budget is being submitted because of a feeling that they look at things too narrowly; that is, they worry about the need to make money today and about expenditures. Financial people can take a broader point of view and can make valuable contributions to new projects. By developing projected profit-and-loss statements and carefully analyzing them, a good financial department can point out those areas where a change will most affect profits. Also, they can point out those items in the projection that, if not met, will have the most adverse effect. In doing this, the newer approaches of mathematical modeling can be most useful. The use of these approaches and the background of financial people can quite frequently lead to insights that are missed by scientific and marketing individuals.

All of my comments above lead to the concept that the creation of balanced teams of individuals with various backgrounds and skills to work on a new project is essential. Creation of a team should ensure that different points of view are heard. and that communications and relationships are improved. Such teams are generally most effective if under the control of one individual designated as project manager or venture manager.

J. T. Baker's development of a suitable grade of manganous carbonate is an example of a successful team project based on close cooperation of commercial development, research and development, and production.

Commercial development saw a need for an improved grade of manganous carbonate in the rapidly growing field of ferrite memory cores in computers. The manganous carbonate being produced and sold by Baker was not entirely suitable. It was difficult to relate electronic performance requirements to chemical and physical characteristics of the product. The Baker research organization was able to identify and characterize suitable products that could be made in existing plant equipment. The customers confirmed the value of the products, and this has been further confirmed by use in computers over the last 6 years.

The development of this project is shown in Table 1. Efforts to build on the success are shown in Table 2.

A few years ago, it appeared that these team efforts would eliminate most of the problems that had previously been experienced in getting effective new product development work done across the many functions of a business. Unfortunately, in many larger companies, they have not been as effective as had been hoped.

I believe that one reason for this in the larger companies has been the fact that individuals serving on a team are not able properly to represent the function

TABLE 1

Successful Project Implemented over Various Functional Lines—Manganous Carbonate Project of J. T. Baker Chemical Company

Action	Functions Implementing
Report manganous carbonate had promise in manufacture of ferrite memory cores for use in computers	Sales
Definition of product requirements; some grades of manganous carbonate appear suitable	Commercial development with prospective users
Definition of potential demand for suitable product in ferrite memory cores	Market research
Experimental effort to identify and characterize suitable grade of manganous carbonate	Research and development (company and customer)
Process design and initial plant production	Production, research and development
Continuing plant production and sales	Production and sales

TABLE 2

Efforts to Ensure Continued Success and Build on Progress–Manganous Carbonate Project of J. T. Baker Chemical Company

Action	Functions Implementing
Redefinition of product to get improved performance	Marketing, research and development, production
Expansion of sales of manganous carbonate into related fields: phosphor uses, thermistors	Commercial development, sales technical service
Study marketing of related products with similar technology: barium carbonate; calcium carbonate; cupric carbonate; magnesium carbonate; nickel carbonate; strontium carbonate	Entire product team
Study related needs in ferrite core market: need for different balance of electrical and magnetic properties to get different performance characteristics in memory cores; need for improved thermal stability of memory cores	Entire product team
Take broader look at needs of the electronic industry	Commercial development

of which they are a part, and creation of the team makes the commercial development person less sensitive to the need for communication broadly to the various functions. Because there is a production person on the team, the need to send information to all the individuals in the production organization that may be involved can be missed. Thus one individual is informed but the vice-president of production or the plant manager or some other individual involved may not be given the necessary background to secure his or her support and needed cooperation. The need for communications to each function has been emphasized. This need exists whether or not teams are formed within the company.

If teams are formed, it is essential that an individual respresenting a function have proper authority within the function to see that the job is done.

A recent article in the *Harvard Business Review* points out that "the beginning of wisdom in talking about the process of technical innovation is recognizing that it is one of the least understood management processes in business." I agree. I offer as a possible prognosis guide typical statements you may hear that indicate if a project is headed toward success or failure (Table 3).

TABLE 3

Project Prognosis Guide

Success	Failure
Management knows of investment need 3 years away and is planning for it.	Management wants new products for needed growth and wants to be filled in on details later.
Research found technology that could be purchased outside that will save 6 months of experimental effort.	Research found three interesting technical leads 4 months after commercial development decided the market is too small to be of interest.
Production asked research to make studies that greatly simplified production in existing equipment.	Total production from the unit hit a record high this month. One small-volume run of a new product was not successful because equipment we have is not suitable.
The sales force can sell the product to existing customers.	Marketing did not know offering the new product would upset the company's best customer.
Development of a mathematical model for the project showed two critical factors needing careful study.	The production unit is able to reach only 75 percent of expected capacity and so the new product may not be as profitable as existing products made using the same equipment.

Although the guidelines suggested here can be helpful in managing new ventures, it should be recognized that there is a greal deal that is not understood as to why certain groups can do so many new things and others have such difficulties. Perhaps this is the reason why we keep coming back to these problems in any discussion of commercial development.

PART IV

OTHER ASPECTS OF COMMERCIAL DEVELOPMENT

14
INTERNATIONAL COMMERCIAL DEVELOPMENT

Robert J. Polacek
Management Supplements
Briarcliff Manor, New York

To the commercial development practitioner whose activities have been limited to the North American continent, the very title of this chapter may conjure up romantic visions; from minarets overlooking an eastern bazaar to Rhine River steamers passing the Lorelei while medieval castles look down from picturesque banks. To the uninitiated, the mystique that surrounds the international aspects of commercial development seems to have been nurtured, consciously or subconsciously, by those who have been exposed to its vagaries and who have clothed themselves in its mantel of expertise, the better to stake out a claim to an area that promises adventure and a relief from dull routine.

This aura of mystery has been enhanced by the surprising dearth of papers and publications dealing with the "how-to" aspects of international commercial development, from which one may conclude that it is either so complicated as to defy description or so simple that any description would be redundant. Actually, the true situation is neither. In this age of McLuhan's "global village," international development is a natural outgrowth of the growing interdependence of the nations of the world.

To the more sophisticated observer, international opportunities and techniques of commercial development have a surprising degree of commonality with those that have been experienced and practiced domestically. Even the obvious differences seem to be more a matter of degree than substance. More companies do not avail themselves of international opportunities because of a fear of the unknown that has little basis in fact. Many of these have successfully developed products in the Canadian market for many years without considering it either difficult or onerous.

If we define a multinational company as one that manufactures and sells in more than one country, we find an increasing segment of U.S. industry matter-of-factly engaging in commercial development in the countries and regions of the world in which they have established operations.

They have found that, with some modifications to accommodate the local business customs and markets, they are able to utilize to a great degree the commercial development techniques that have been developed domestically for many years.

It is for the large number of companies that, through timidity or lack of information, have ignored the potential of international markets that this chapter will be of greatest value.

There are many reasons why a company should consider taking advantage of international marketing opportunities. Not the least of these is the fact that international markets afford an opportunity for the technological enterprise to use existing technology to obtain additional profits in new geographical markets without incurring substantial additional research costs. Opening international markets for existing products has the effect of multiplying the profit return from investment in research and development that, at least in a well-managed company, has already been paid for from domestic sales revenues.

Because no country is self-sufficient from the standpoint of technology, exposure to the needs of international customers and to international competitor's products opens the door to a two-way exchange of technology that can benefit a domestic company in its U.S. marketing efforts.

Finally, the expansion of trade in products and technology is vital to the long-term interests of the world and of the United States. The endemic U.S. balance-of-payments problem, which has repercussions throughout the industrialized world, could be significantly alleviated if more than the current 10 percent of the 300,000 U.S. manufacturing enterprises were to take advantage of international markets.

There are various alternatives from which a company may choose in order to expand its interests internationally. Although many variations exist, the basic alternatives are four: export sales, licensing technology, establishing a joint venture, and forming a wholly owned foreign subsidiary. These alternative means of participation in international markets are listed in increasing order of degree of financial risk and also in increasing order of financial return. One of the principal functions of any international commercial development activity is to select the alternative that best satisfies the requirements of the situation. The choice will depend on the degree of sophistication of the company, the economics of the situation as dictated by the foreign competition, raw material prices and availability, freight considerations, the physical form of the product, chemical characteristics, foreign exchange rates, and the peculiar characterisitcs of the market in the country or region in which the product is to be sold.

Export Sales

Overseas selling is generally the easiest way to begin developing an international market and in most cases should be resorted to as an initial step, even if the ultimate objective is to produce the product in a foreign country.

Frequently, the solely domestic-oriented company begins to export by default rather than as a planned strategy for growth. An inquiry for the product is received, samples and literature are sent, the product is accepted, and an order placed by the foreign company. Usually, the domestic supplier will work through an export agent or broker to consummate the trade. Considerable assistance is available from the U.S. Department of Commerce and the larger U.S. banks whose staffs are familiar with all phases of export trade. The U.S. Department of Commerce, Bureau of International Commerce, maintains field offices in many major American cities and can be extremely helpful in advising on how to handle a company's exports to various parts of the world.

By a concerted follow-up effort, it is possible to expand this initial export sale by first determining the reasons that prompted the customer to purchase the particular product and then contacting other potential customers. Expansion is usually accomplished most effectively by establishing agents in the various countries in which the product has a likely market.

If the product is relatively unique in performance and has a sufficiently large profit margin, the export sales strategy should be pursued. The next step is to establish the market potential and an agent distributor network, which can be done by using internal personnel or marketing consultants familiar with, or actually located, in the target areas.

The usual preliminary information regarding size of market, prices, competition, technical service requirements, and sales promotion techniques is necessary. In order to determine the profitability, additional information is required regarding ocean freight rates, insurance, tariff rates, inland freight rates at the destination, foreign taxes, and government regulations regarding product use and sale. Most important, the actual application requirements for the product must be definitely established. Many U.S. companies have established substantial and profitable domestic sales for their product, only to find that the international performance requirements or methods of application differ a great deal, requiring sizable substitute product modification or even precluding the product use entirely.

Warren Keegan enumerates several possible strategies that can be used to adapt a product to the differing conditions of foreign markets.[1] These range from simple product extension by which the unmodified product is sold overseas in the same form and with essentially the same marketing techniques as in the United States to the creation of an entirely new product to meet the specific needs of other countries. The outstanding success of Coca-Cola is an example of the simple extension, whereas the development of new low-cost, high-protein supplements would be an example of the invention strategy. Obviously, the costs and risks of the simple extension are considerably less than those involved in the product invention approach.

Between these two extremes is the strategy of marketing adaptation in which the marketing of the unmodified product is shaped to accommodate differing local needs; the strategy of product adaptation that involves additional product modification so that it more exatly fits local customs and needs; and finally, the strategy of dual adaptation in which both product and marketing are changed to enhance the product's acceptability. These mid-spectrum strategies are cited in their increasing order of cost and risk involved.

Marketing Channels

The development of the marketing channel by which to enter the foreign market is a key factor determining the success of the international operation.

For the firm just beginning to look toward the international market, and without an established set of contacts abroad, the indirect method is probably the easiest and least costly method. This involves selling to a domestic-based exporter who purchases the product in the United States and sells on his own account or to an export agent or broker who finds foreign buyers on a commission basis.

The direct method involves the selection and appointment of distributors and agents in the operational area. Although this approach is more difficult and involves a greater level of risk, it is usually more profitable. The direct approach provides a greater facility in exploiting opportunities due to a higher level of control and improved communication between the manufacturer and the ultimate consumer.

Remember that selection of foreign agents should involve at least as much care as selection of domestic agents. It is well to keep in mind, however, that none of the alternative choices is likely to be perfect. Continual search for the ideal may result in complete inaction. Allow a reasonable length of time for the search and choose the best from several alternatives, otherwise you may, like the Minotaur, wander the labyrinthian maze forever.

The question as to how many agents are needed is one that can be answered only in light of the existing situation. Generally, it is wise not to be lulled into a misleading sense of unity when considering the Common Market. In spite of the significant progress in standardizing commercial law, lowering of tariff barriers, and the like, each European country remains a separate and distinct market. Although many agents have sales operations in more than one European country, any reasonable investigation will reveal that the agent does best in one or at most two market areas and relatively poorly when over-extended. It is well to keep in mind that both industrial and household consumer tastes and practices may vary widely.

The obvious conclusion then is that a top-flight agent should be chosen for each major market area.

The newcomer to the field may find some unexpected difficulty arising in his or her search for a suitable agent. Predatory predecessors who have had export sales agreements, only to terminate them once the foreign agents have built up a sufficient volume to warrant the exporter's establishing its own sales forces, have left a residue of ill will toward U.S. companies. Although it is not my purpose to pass judgment, I would suggest that a liberal application of the golden rule would go far in alleviating the image of the "Ugly American corporation" using its foreign distributors. If the eventual establishment of the U.S. company's own sales force is an objective that has been predecided, the foreign agent should be told of the decision and given sufficient time and compensation to effect an orderly and rancorless transition.

If we were to choose one simple term that is the most important key factor for success in establishing a mutually profitable relationship and growth in export sales through a foreign distributor, that term would have to be "communication." Many U.S. firms spend considerable time in selecting an agent and negotiating a satisfactory contract and then expect everything to progress of its own momentum. What momentum? Very little has been established. If you haven't established forward momentum toward the top of an inclined plane, retrogression should not be surprising.

Written reports, new applications data, personal visits, and exchange of significant call reports are essential in developing and maintaining a growing sales situation. Help your agents to sell your products by establishing routine channels of communication. Assist the agent in translating your technical data sheets into the language of his or her country. Remember that these translations should not be too literal. Little things such as units of measurement should be changed to conform to local practices.

Advertising reprints and other promotional literature should never be translated for use in foreign countries, even with modifications. Remember that you are trying to influence people with different cultural associations. Acronyms and slang are particularly dangerous to employ—what may provide "zip" to a U.S. promotional piece may provide "zap" in a foreign country.

Occasional publication of a technical article by the exporting company in a reputable foreign journal is helpful in promoting an image of more than an absentee supplier. By using co-authorship, you may also identify your agents with the product. Remember that export sales must be promoted and supported just as well as your domestic sales.

Trade with the Far East, and in particular with Japan, is unique in its extensive use of Trading Companies. Much confusion results when the noninitiated confuses these export-import operations with agents or distributors such as we know them in the United States. Trading Companies are generalists in export-import trade with excellent connections with all segments of Japanese industry. They do not specialize for the most part and do best with large-volume commodities. Usually not very promotion minded, they may not be the best choice for specialties that may require considerable technical service or promotion. They frequently handle competing lines. Nevertheless, for those not familiar with the vagaries of the Japanese distribution system, Trading Companies may be the easiest way to start. Finding agents or distributors before becoming familiar with the system can be frustrating.

As mentioned previously, the U.S. government is encouraging export trade and has provided a simple way to defer taxes on profits derived from foreign sales through the formation of an export subsidiary known as a Domestic International Sales Company (DISC). In spite of the seeming contradiction in nomenclature, it would be to most companies' advantage to look into this possibility.

Finally, it is important to make certain that the method of payment is agreed upon in advance with the agent. It is well to avoid the more complex transactions such as consignments, and so forth, until you have sufficient experience with a particular agent. The irrevocable letter of credit is probably the best way to get started. If credit is a problem, it is better to compensate for immediate payment by increasing the sales discount than to enter into complex credit arrangements.

Licensing

There are many reasons why it may not be profitable to export products. The more obvious physical reasons are toxicity, lack of stability, flammability, and physical form (a 50 percent active agent polymer emulsion would necessitate shipping water over distances that would be prohibitive). Tariff rates and high freight costs could price the product out of the market.

In these cases, licensing of technology can be a profitable alternative. Indeed, international licensing can be a significant source of earnings for products and processes even if the inhibiting factors to export as described above are not relevant. In the late 1960s, ICI reported an income from licensing of approximately \$38 million. A medium-sized company can utilize the licensing of its products to expand its product line through an exchange of technology, or cross-licensing. The resultant savings in development costs can be substantial. But product applications can differ, and research on modifying products or developing new applications data on products obtained through licensing in order to meet domestic requirements can also be substantial.

There are several salient considerations that must be taken into account in licensing technology. Not the least of these is the factor of introducing new competition into the marketplace. It certainly behooves the licensor to choose its timing so that it has already established a strong market position in its own traditional market area before it proliferates its technological knowledge.

A second important consideration that frequently traumatizes the unsophisticated potential licensor into inaction is that of secrecy. The more people who have access to the technology, the greater the risk of leakage. A compensating factor is the fact that most employees of foreign companies do not enjoy the job mobility of the U.S. employee. It is highly unusual in most parts of Europe and Japan for a person to change jobs, even more unusual for a change to a competitor. Also, although limited in effectiveness by some local laws and customers, it is frequently possible to provide additional protection in the license agreement by making secrecy agreements mandatory for all of the licensee's employees who will receive confidential information. This, of course, places an added burden on the licensor to adequately designate the truly confidential information, so as not to inhibit the proper implementation of the information and prevent adequate application information from flowing to the customer. As in most factors dealing with international relationships, common sense and an understanding of the foreign national's psychological makeup is an essential ingredient to a satisfactory agreement.

A third key consideration in licensing technology is whether or not a patent has been issued covering that technology. Usually, an unsophisticated U.S. company has not taken the trouble to establish a patent position in the more important industrialized countries of the world. This is often a mistake. A patented product gives a firm a significant negotiating edge. The problem, however, is to determine in which countries a product patent should be obtained. Nowhere does the lack of cohesiveness in the EEC exhibit itself as dramatically as in the field of patents. Although most of the industrialized nations of the world are members of the International Convention for the Protection of Industrial Property, each has its own patent code. Obviously, one cannot patent a product everywhere—the cost would be prohibitive. Discriminating selection of the countries in which the technology would be most likely to have a significant impact is a necessary chore. There are patent firms that specialize in helping U.S. companies get and, equally important, maintain patents in foreign countries. It is well to establish a relationship with one of these legal firms if you are going to take advantage of foreign markets for your technology.

Foreign patents can be used to protect exports by the patent owner to that country and allow the exporter much more flexibility in negotiating agreements for its distributor. For example, a patent gives the exporter the legal right to limit the patented product's use, sale, and manufacture, to restrict the sale to a particular use or a specific country, and to limit the term of the agreement to less than the full life of the patent. In the absence of a patent, any such restrictions may be invalidated by national or EEC laws regarding retraint of trade.

There are other potential legal pitfalls that should be avoided through the prudent use of legal advisors. There are many multinational legal firms practicing in the United States that have branch offices in one or more major countries overseas. Most of these specialize in the commercial laws of specific geographic areas—for example, the EEC, Japan, or Latin America—and can be of considerable assistance in helping to get the job done with a minimum of legal risk.

A fourth consideration in licensing technology is the need for continuing communication. The requirement for initial transmital of the technology package and training of the licensee's personnel is generally well understood. The question of what type of continuing know-how should be transmitted and how it should be transmitted is much more difficult. Because the royalty income to the licensor should be directly proportionate to the success of the licensee with the product or products in question, in most cases it behooves the licensor to provide continuing know-how related to improved technology and marketing. If the licensee is proficient in the particular product line or marketing area, there should normally be a provision for a back-flow of information to the licensor. Any improvements to the technology should normally accrue to the licensor on a royalty-free, and, if necessary for agreement, on a nonexclusive basis. Wherever possible, the right to sublicense these improvements to other licensees should be maintained. It would appear that, if the licensor and all of its licensees can share in the best improvements of each, a more vigorous and healthy situation will exist for all concerned.

The manner in which the communication loop is maintained is most important. First of all, although we normally license technology of manufacture, it is equally important to supply the technology for marketing—that is, applications information, market informations, and promotional ideas. Obviously, not all will be pertinent in every part of the world. Also, sending frequent and voluminous reports is onerous to both the sender and the receiver in that the expenditure of time frequently exceeds the value obtained. A most effective means of communication is for a periodic exchange of visits during which the licensee can spend time in the licensor's plant and laboratory and in the field to find out what is new and what part is applicable to his or her needs. The same holds true for the licensor. If visits are arranged on a quarterly basis, it means that each party will be the host twice a year. Experience has shown this to be the most cost-effective technique where the host has done sufficient prior planning to make the visit meaningful.

The Joint Venture

The next step in the hierarchy of international involvement is a joint venture to produce and sell products internationally. The joint venture has the advantage of sharing the financial risk with a partner who is indigenous to the area and is knowledgeable of the laws, customs, and market needs, as well as the political, social, and economic situation, in the area in which the joint venture is to function. Although sharing the risk in a joint venture has its merits, the financial rewards must also be shared.

In addition, it should be recognized that finding a partner willing to share the risk does not perforce mean that the risk is an acceptable one. The partner may assess the risk differently. He may be willing to take a greater risk than the domestic company would normally be willing to take, and his judgment as to the extent of the reward that would be acceptable may be different from that of his partner.

Many companies lose sight of the fact that a joint venture requires a greater involvement than that of a licensing arrangement. As much care should be taken in the choice of a joint venture partner as should be taken in the choice of a marriage partner. Each partner should have the capability of contributing to the joint venture in a measure commensurate with his share in the equity. Management philosophies, strategies, and objectives should be compatible. The information on the market, competition, costs, government regulations, labor practices, taxes, and so on, should be at least as detailed as would be required for a domestic investment. A 1- and 5-year pro forma profit-and-loss balance sheet and a cash flow study indicating the proper amount of capitalization should be prepared, studied, and thoroughly understood by both parties.

Finally, no joint venture should be attempted unless each of the partners has a surplus of managerial resources to devote to the venture. If this is not the case, the licensing or export routes are more prudent alternatives.

Whenever possible, the joint venture should be preceded by a period of courtship during which the potential partner acts as a sales agent for the products that are exported from the United States. During this period, each potential partner should gain an insight into the actual capabilities of the other, and both will be able to develop and refine the information needed on which to base a valid investment decision.

Once the joint venture has been established as the best alternatives, and the joint venture partner is selected, the question arises as to the type of agreement that should be negotiated. Although some companies prefer a simple agreement outlining the venture in broad terms and relying on a spirit of cooperation and good faith to resolve any differences that arise, others prefer a detailed agreement covering as many contingencies as can be foreseen. A detailed agreement need not indicate a lack of trust but rather serves to identify those areas in which differences of style, opinion, law, and custom exist between the partners. In most cases, these differences can be compromised to the satisfaction of both partners during negotiations. Where the differences are irreconcilable, the detailed exchange of views at least allows management to decide clearly whether they are important enough to jeopardize the validity of the venture. It is important in a joint venture to know what you are getting into before you take the step, to know where the philosophical and strategic differences exist and to be prepared to accept those whose acceptance may be necessary for the effective functioning of the enterprise.

In any event, a good agreement should have built into it sufficient flexibility for change when the situation warrants. It is never possible to foresee all of the problems that arise, but by anticipating some of them and negotiating a mutually acceptable solution through the basic agreement, the precedent has been set for subsequent give-and-take, which will very likely solve most, if not all, of the ensuing problems.

The Wholly Owned Subsidiary

The establishment of a wholly owned subsidiary to produce and sell in a foreign country is normally considered the riskiest method of participating in the international market area. There is considerable validity, however, to the argument that this method involves less of a risk than the joint venture approach because the problem of accommodation to the wishes of a partner is not a factor.

I tend to lean toward the latter assessment, particularly if the risks have been minimized as described in the following paragraphs, if sufficient capital resources are available or can be made available to the domestic company, and provided the host country allows a wholly owned foreign operation.

The primary method of reducing the risk is to have accumulated some experience in the target country or area through a vigorous, properly supported export program. The same detailed marketing, commercial, and financial information that should be obtained for a joint venture will be required for the establishment of a wholly owned subsidiary operation. The existing agent network can continue to be utilized, and its effectiveness should be improved with local manufacturing and technical service support. Most likely the level of sales that has been established will have provided the justification for local production.

One of the least expected benefits of establishing a foreign manufacturing operation is that it frequently results in an increase in the export volume of the domestic parent by providing stimulus to the sales of other domestically produced products. The record of joint ventures in this regard is not quite as good.

It is, of course, most important to have competent and adequate legal advice when setting up the foreign corporation. As in the case of the joint venture, consultation with a multinational legal firm knowledgeable in the laws of both the United States and the target country is the best approach.

Probably the most difficult aspect of establishing the foreign subsidiary is the selection of the personnel to operate it. Most successful ventures begin with a U.S. national in charge assisted by a deputy who is native to the country and who will be expected to assume total responsibility for the operation as soon as possible after the entrprise has been established. Part of this individual's training must include a suitable period of time at the parent company to enable him or her to become adequately acquainted with the parent's philosophy of operation, its support capabilities, and the key personnel on whom he or she will depend for support.

Although a wholly owned subsidiary frequently gets more support than a joint venture, licensee, or sales agent network, there is no logical reason why this should be so, other than the fact that the parent has a greater financial stake in the wholly owned operation. The extent of the success of any of these approaches is directly proportionate to the amount of support that is given.

New Product Testing

There is an increasing use of foreign markets as an aid to domestic commercial development activities. Increased U.S. government regulation has drastically extended the time it takes to position products on the U.S. market. This is particularly true in the areas of pharmaceuticals and pesticides, where even preliminary market testing requires FDA and/or EPA approval. For those companies that have become familiar with select international markets, it is frequently possible to obtain the necessary foreign approvals and begin test marketing months, and even years, before it would be possible to do so in the United States. In fact, the results of these foreign market tests, provided they are properly controlled and documented, are frequently useful in helping to obtain approval by the U.S. government regulating agency.

There can be other advantages to preliminary market testing in foreign markets. There are many examples. Agricultural chemicals can achieve accelerated testing by taking advantage of the multiple growing seasons in many parts of the world. Unencumbered by overlapping and frequently confusing governmental building codes, chemicals and plastics that have potential application in construction can often be tested and developed with a minimum of bureaucratic delay in many parts of the world. Protective coatings can be tested under conditions of extreme climatic conditions. Product "bugs" can be worked out in discrete markets without endangering the product's image or reputation in markets representing a large potential. Finally, there may be innovative products for which there is little need in the United States or other highly industrialized markets, but they may be the kind of products that will contribute to improving the quality of life in the less developed countries.

International commercial development of this nature would seem to pose few new problems not encountered in domestic commercial development activities. Although there are adjustments that must be made to comply with local conditions, the principles of successful commercial development will generally be the same.

Conclusion

International marketing uses the same basic marketing tools, concepts, techniques, and knowledge that are used in domestic marketing. There can be, and usually are, differences in consumer attitudes, in institutional practices, and in political, sociological, and cultural factors.[2]

For those who have had little experience in foreign operations, several general principles should be followed. Foremost among these is the fact that each foreign country is a separate situation offering both unique problems and opportunities. Do not be misled by assuming that common market areas are homogeneous groups—they are not.

Secondly, a U.S. company doing business in foreign countries must adapt to the customs, cultures, and laws of that country. The converse is unlikely to occur. What is best or what works in the United States is not always the way to success in another country. Indeed, it frequently is not. It is somewhat strange that most U.S. companies will easily adapt to regional U.S. differences, varying their approach on the West (East) coast, in the South, and in New England, but seem to lack this same flexibility in dealing in a foreign country.

Thirdly, the basic principles underlying the various development techniques are essentially identical whether applied domestically or internationally; only the method of application is likely to differ depending on the differences in the culture and business practices that are encountered.

Finally, although the profit motive is and should be the principal driving force in developing international operations, it does not preclude satisfaction in knowing that these operations, properly handled, will further the national interest by contributing to the balance of payments, opening up communications at a business and technical level with our fellow inhabitants of the globe, and, in some cases, supporting the effort of the less developed countries to provide a higher standard of living for their citizens. Improving profitability through international operations can provide both exciting opportunities and satisfying challenges.

Notes

1. Warren J. Keegan, "Multinational Product Planning: Strategic Alternatives," *Journal of Marketing*, January 1969, pp. 58-62.
2. Eugene J. Kelley, and William Lazer, *Managerial Marketing: Policies, Strategies and Decisions* (Homewood, Ill.: Richard D. Irwin, 1975), p. 312.

Bibliography

Avon, P. H. "Transmission of Technical Information on the Licensing Framework." *Papers Presented at ACS Mtg., Division of CM&E,* New York, September 1966.

Carpenter, W. Sam. "Challenges in World Trade." *Papers Presented at ACS Mtg., Division of CM&E,* New York, September 1963.

Corwin, R. C., Hedley, J. M., and McEwen, R. L. "European Chemical Market and Economic Information: Problems, Approaches and Sources." *Papers Presented at ACS Mtg., Division of CM&E,* Philadelphia, April 1964.

Graziano, Frank J. "Commercial Development Abroad to Maximize Profits at Home." *Changing Times in Commercial Development: Papers Presented at CDA Meeting,* New York, March 1971.

Kain, Walter. *Acitivities of American Chemical Companies in Europe.* Baden Baden, West Germany, 1963.

Keegan, Warren J. "Multinational Product Planning: Strategic Alternatives." *Journal of Marketing,* January 1969.

Kelley, Eugene J., and Lazer, William. *Managerial Marketing: Policies Strategies & Decisions.* Homewood, Ill: Richard D. Irwin, 1975, Chapter 3.

Kenen, Peter B. *International Economics,* 2nd ed. Englewood Cliffs, N.J.: Prentice-Hall, 1967.

Kischitz, A. D., ed. *An International View of the Industrial Commercial Development: Papers Presented at CDA Meeting,* Williamsburg, Va., May 1973.

McGann, P. W. "Issues Concerning Foreign Trade in Chemicals." *Papers Presented at ACS Mtg., Division of CM&E,* New York, September 1966.

Morgan Guaranty Trust Co. *Export and Import Procedures,* 3rd ed. New York: International Banking Division of Morgan Guaranty Trust Co., 1963.

Sneath, William S. "Financial Considerations of International Trade." *Papers Presented at ACS Mtg., Division of CM&E,* New York, September 1963.

Stanley, A. O. *Handbook of International Marketing.* New York: McGraw-Hill, 1963.

U.S. Department of Commerce. *Exports: Growth Market of the 70's. New York Times,* September 30, 1973.

Wade, W. "Legal and Patent Problems in Import-Export Marketing of Chemical Products." *Papers Presented at ACS Mtg., Division of CM&E,* Philadelphia, April 1964.

15
FUTURE TRENDS IN COMMERCIAL DEVELOPMENT

Newman H. Giragosian
Delphi Marketing Services, Inc.
New York, New York

Robert J. Polacek
Management Supplements
Briarcliff Manor, New York

Previous chapters have described the changes that have taken place in the concept and practice of commercial development. It is apparent that several patterns are emerging.

First, the formalized structure of a commercial development department that was common the 1950s has been superseded in all but a few companies. The need for commercial development as a function, however, has become even greater than in the past. Whether or not other organization structures are found that will perform the process of commercial development in the future is really irrelevant; what is pertinent is that commercial development is not only a necessary and vital activity essential to the profitable growth of a company but that it is perceived as a vital function necessary to the profitable growth of a company.

The growing complexity of the future will require that commercial development personnel be knowledgeable in the areas of distribution, operation research, marketing global markets, among others.

The coming decade will witness more societal control of chemicals and how they are used. Regulatory agencies may be expected to be even more zealous than the courts in discharging their statutory obligations. Generally, lawsuits involving product liability will increase, with manufacturers more vulnerable in cases of negligence in design or lack of reasonable foresight. Warranty disclaimers and privity of contract (the concept that an injured plaintiff cannot hold liable a manufacturer with whom there is no direct contractual relationship) will afford the defendant less protection. In the absence of countertrends, liability insurance may be more costly, and companies will be faced with greater risks.

Legislative and regulatory influences deriving from ecological considerations and consumer interests will also make life more challenging. Even when government regulation is avoided by responsive industry action, the logistics of adjustment are frequently staggering. Consumer interest will become even more important. While industry must cater to consumer interest for a variety of good reasons, government increasingly concerns itself with protecting the consumer from personal hazard due to product design. However, the consequences of consumerism as a force for change are by no means negative. Concern over cigarettes, for example, has created a new market for filters and adsorbents, whereas safety belts in automobiles represent a new market for synthetic fibers.

The impact of business trends will also have a great effect on commercial development. The acceptance of innovation and change as business style, the need to maintain high levels of production and high rates of growth in industry, the complex and competitive background against which crucial business decisions must be made, all create a climate in which the work of commercial development will become even more critical than in the past. Because some companies are willing to take greater risks than are others, the time to move a new product from concept to market is diminishing.

Changing Nature of Commercial Development

In response to the changing forces affecting the chemical industry, commercial development has changed not only in character but also in the way in which it carries out its traditional functions. Some of the forces affecting the chemical industry, and in turn commercial development are:

1. The changing nature of the chemical industry itself—that is, the blurring of lines of distinction between the chemical industry and the industries it serves
2. The impact of changing process technology
3. The increased emphasis on international operations
4. The trend of chemical companies toward vertical integration
5. The increased trend toward mergers and acquisitions
6. The increased difficulty of generating sales per dollar of research and development expenditures

To say that the chemical industry has been grounded in research is a truism. Beginning in the 1930s, the U.S. chemical industry began to change from a research-oriented group to a production-oriented group, and still later to a marketing-oriented industry. As the boundaries of the industry were extended, the importance of finding markets for the many new products conceived by the industry became more important. Inevitably, with the proliferation of new products and the maturing of older products, process technology has had more impact on chemical markets. Since the 1950s, the chemical industry has become a truly worldwide industry. Not only have U.S. exports been increasing each year, but also U.S. companies have invested heavily in plants abroad. This has occurred partly as a result of action by the Common Market, which has raised tariffs to companies not having plants within the Common Market, but also because U.S. companies have found it an advantageous method of reinvesting money earned in foreign countries. The total commitment of U.S. companies in foreign countries in 1978 is estimated at perhaps $10 billion, and foreign earnings of U.S. companies are extremely large.

As a result of this increased investment abroad, and of the growing importance of foreign earnings, it is becoming increasingly important for companies to orient their commercial development activities accordingly. Knowledge of foreign competitors and their marketing capabilities is also important. At the same time the impact of foreign producers on U.S. domestic markets has made it imperative for U.S. companies to learn more about the capabilities of their foreign competitors from the viewpoint of their ability to compete successfully in U.S. markets.

One of the more prominent and publicized marketing trends of the last 20 years has been the forward integration of chemical companies toward the consumer. By forward integration is meant the sale by chemical companies of consumer "shelf" products, such as hair sprays, cosmetics, and automotive chemicals. There are, of course, degrees of forward integration, and a basic raw materials producer may feel that it has integrated forward if it simply upgrades the raw materials into intermediates and makes some progress down the line of distribution of material to the ultimate consumer.

As companies move forward to the consumer, further challenges will face the commercial chemical development practitioner, who up to now has been primarily industry oriented.

Companies have increasingly been attempting to maintain growth in sales and profits by acquiring other companies in related, as well as nonrelated, fields. Commercial development has become the spearhead for this type of growth.

The time allotted for commercialization of new products in the chemical industry has been decreasing significantly. Competitors are moving in more rapidly than in the past, and commercial timing is more important to achieve economic success.

For each dollar of research and development expended, the sales dollars that have been generated from these developments have been decreasing.

All these changes in the chemical industry have forced changes in the emphasis within the commercial development function away from the market development aspects of internally generated research products, as was the case in former years, to a broader concept of the commercial development function and a recognition of the critical nature of the timing of commercial development.

Already during the 1960s and 1970s we have seen the need for a more considered approach for planning of development and new product ventures. Let us now turn to the future to see what are the challenges facing the chemical industry in the planning of development and new product ventures.

Challenges of the Future

Increasing Rate of Technological Change

The rate of technological change will continue to accelerate, which will create more pressure on companies to keep abreast of technological developments.

Increasingly, the marketing research groups have been doing an excellent job of carrying out sophisticated marketing research sutdies on problems involving current product lines or products, while new to the company or at least current in the market place, or have substitutes that are currently recognizable. Looking forward to the future, however, it is easy to see that with the increasing number of technological breakthroughs, marketing research groups must be able to do imaginative new product research. Manufacturers of industrial products today are looking for new ways to minimize the risks inherent in the development and marketing of new products. They are emphasizing more and more advance research in pretesting new product ideas.

Increasing Economies of Scale

This means the acceleration of the current trend toward much larger chemical plants. We see today that many companies are turning increasingly in the direction of building huge plants regardless of the ability of the market place to absorb the possible production from these new complexes. If companies find that their unit cost for a certain product will be substantially reduced by a much larger facility, they appear to be increasingly willing to build the larger complex and operate the plant at a lower percentage of capacity, hoping that demand will eventually catch up.

Increasing Cost of Research and Development

Sales per dollar of research expenditures in the chemical industry have been declining steadily. This trend is expected to continue with even greater force during the next decade because of two factors: (1) the rising cost of research per se and (2) the growing difficulties of developing new products superior to those already in the market place. This will put more emphasis on new product research to aid in research guidance. This has already begun in a rather small way but will grow in importance as part of the responsibilities of marketing research departments.

Growth by Merger and Acquisition

During the past 20 years the chemical industry, in addition to relying upon internal growth via the research route, has increasingly chosen to grow by acquisition and merger. With the growing absolute and relative cost of developing new products by internal research, we expect companies to continue to look for growth opportunities via acquisitions. In responding to this trend, the nature of commercial development will change. During the 1950s it was essentially a market development function, in which products discovered in the laboratory were then developed commercially by the market development or product development group. During the 1970s and 1980s we expect to see implemented a broadened concept of commercial development. The commercial development function will probably consist of several different groups, all designed to achieve the same end—namely, successful commercial development of new products and planning for company growth by whatever route seems most feasible.

Lower Profit Margins

We are all familiar with the declining profit margins experienced in the chemical industry during the past 15 years. This has occurred in part because of the aforementioned forces and in part because of increasing international competition. Few people realize, however, that the chemical industry is in danger of losing its traditional image of being a growth industry. As a matter of fact, it is behaving more and more like a cyclical one.

During the early 1950s, profit margins for chemical and allied products grew steadily in comparison with all manufacturing, but since the late 1950s, margins have declined and have been lower than profit margins for all manufacturing. Similarly, profit margins as a percent of gross plant for the chemical industry have been declining as compared with all manufacturing.

It has also become more difficult to generate sales for a given dollar of plant expenditure. This is a problem shared with other industries.

This is not to say that there are no profit opportunities left in the chemical industry. We all know that certain portions of the chemical industry are growing faster than others, and it is really more useful to talk about the chemical indusry as consisting of perhaps 35 or 40 separate businesses than being a homogeneous industry. There are a number of growth areas remaining to be explored, and the race in the coming decade will be won by those companies capable of identifying profitable growth areas. It will be an opportunity for a sophisticated, strong, and imaginative commercial development group to contribute to the growth of the company.

PART V

SELECTED CASE HISTORIES OF COMMERCIAL DEVELOPMENT PROJECTS

As pointed out in Chapter 3, "Innovation and Creation of New Product Ideas," a new commercial development project originates from the discovery of a new product to fill a perceived need. This chapter illustrates this point. Several case histories have been prepared that illustrate how a project evolves. These cases are:

1. Adhesives–Loctite Corp.
2. Corfam–E. I. Du Pont de Nemours & Co.
3. Silica pigments–PPG Industries
4. Noryl resins–General Electric Company

Case History: Loctite Corporation–Adhesives

Introduction

Loctite Corporation today is a multinational manufacturer of adhesive specialties. The company pioneered the development of adhesives to secure nuts and bolts against vibration. This first product was a one-part "anaerobic" adhesive with the unique capability of curing in the *absence* of air, eliminating the need for mixing or heat-curing.

Today the company has over 40 varieties of sealants and adhesives with a broad range of properties. It has developed adhesives for locking and sealing threaded assemblies; for mounting bearings, gears, and other parts; and for

*A number of other case histories have been presented at various symposia conducted by the Commercial Development Association. References to these are found in the Bibliography at the end of this chapter.

sealing flanges without gaskets. It produces preapplied coatings for locking threaded parts, and a "Pipe Sealant with Teflon" for sealing threaded joints. And it manufactures structural adhesives for bonding flat parts.

Founded in 1953, the company has averaged 25 percent growth per year. As of the fiscal year ended June 30, 1978, sales were $127,900,000 and net earnings reached a record $15,536,000.

The story of the commercial development of the "anaerobic sealants" begins a number of years ago.

Anaerobics Discovered

In 1945 at the General Electric Research Laboratory, Birger W. Nordlander and Robert E. Burnett discovered a substance they called "Anaerobic Permafil." It was composed of tetraethylene glycol dimethacrylate monomer carefully oxygenated at elevated temperatures and then stored with a stream of air bubbling through it. A drop of the material placed between crossed microscope slides or metal strips rapidly polymerized at room temperature, thereby bonding the two parts together. Nordlander and Burnett applied the word "anaerobic" to this behavior, indicating that "it came to life in the absence of air." The basic monomer was marketed for a number of years by General Electric Company along with instructions on how to aerate the material in a chemical reaction kettle to produce Anaerobic Permafil. The final product could not be used as an article of commerce because it was tied by an umbilical cord to an air compressor. Nevertheless, a number of customers were developed who used it regularly in production for jobs that included locking of threaded parts but more usually in the sealing of joints and cracks against leakage. In 1952, the General Electric Chemical Materials Department dropped this product from its line, transferring what little business existed to an alternative supplier of the monomer at no charge, with the objective of simply avoiding any harm to these customers. Although Anaerobic Permafil was patented, it was not deemed by either donor or receiver to be a matter of sufficient importance to bother with the formality of a license to be included in the transfer of this business.

Anaerobics Brought Under Control

Although the material failed to achieve general use, Nordlander and Burnett had discovered a material with extraordinary properties. When Dr. Vernon K. Krieble, head of the chemistry department of Trinity College in Hartford, heard of the situation from his son Robert, then manager of chemical development for GE, he was intrigued. He believed that the answer to the problem was to increase the material's sensitivity to the inhibitory effect of oxygen by 1000 fold. One of the things he tried was to oxgenate a similar molecule without the polymerizable double bond, namely–tetraethylene glycol dimethyl ether. Indeed, when this was mixed with TEGMA, he got an anaerobic cure and discovered that the mixture could be stored for years in polyethylene bottles because they are permeable to oxygen. Thus was born the first commercially practical anaerobic sealant that could be shipped fully prepared as an article of commerce. It required neither chemical processing nor even mixing on the part of the customer. The world had a new and far more convenient adhesive. Here was a liquid that stored for years in the bottle but that cured to a permanent plastic simply by confining it between closely fitted parts without need for heat or mixing. It could be as thin as kerosene and highly capillary-active. It would wick into joints spontaneously and, once removed from contact with air, it would harden automatically.

What's It Good For?

The substance was truly unique. However, it was not without some rather troubling properties. The so-called adhesive was brittle and wouldn't adhere to everything. Who might use such a product?

One of many possible applications listed in the Nordlander patent was the locking of set screws. The new anaerobic composition proved to be very well suited for this application and, in fact, for nuts and bolts and threaded fasteners generally. Much data were gathered on the performance in locking nuts to bolts. The Battelle Memorial Institute was brought in to make a market study. They dutifully reported a substantial market for thread-locking devices but concluded that a successful product must be reusable and recommended that research be continued until the feature of reusability was added before trying to commercialize the product. A most valuable recommendation in their report was to price it competitively to lock washers, even though that represented a very high price per pound.

Decision to Commercialize

In spite of Battelle's advice, Dr. Krieble and his associates felt that automotive equipment manufacturers were not so concerned with the reusability of thread-locking devices. And in the aftermarket and the consumer field, reusability would be of no consequence. They decided to commercialize the product. The American Sealants Company was formed with Dr. Krieble as president, and a pilot plant was set up near Hartford. Tests were run for about a year in central Connecticut. The first sales manager introduced the product to local manufacturers, and a year later, a dozen of them were using it regularly in production and were well satisfied.

Initial Approach

In July 1956, the younger Krieble (who now heads the company) left his job at General Electric and joined American Sealants. Based on the results of the first users in Connecticut, the company decided to focus on thread locking in the OEM market through manufacturers' representatives. It was recognized, however, that the Loctite adhesive was (1) too thin and strong, and (2) too unreliable in cure, depending on the surface condition of the parts, which accelerates the breakdown of the hydroperoxide initiator. Accordingly, research was done in developing weakeners, thickeners, and a primer containing additional catalysts. The result was a test kit containing a strong grade, a mild grade, a thick grade, and a bottle of primer to assure cure regardless of surface condition.

After pinpointing the market area and distribution channels, the final decision was in regard to pricing. Confirming Battelle's recommendation—price to lock washer value—calculations showed the product was worth a great many times its production cost. But this fact had to be communicated, and users had to be educated to the benefits of the new technology. The company thus decided that some of the cost savings would be passed on to the users but that some would be used for communication. A 10-cubic-centimeter bottle was priced at $4.50 with quantity discounts running up to 50 percent.

Product Officially Announced

The product was announced at a press party at the University Club, New York City, in July 1956. The presentation explained the seriousness of the problem of loose screws, how they figured as a cause of machinery breakdowns, how lock washers could not completely solve the problem, and how the new company's mission was to eliminate lock washers because they had something that worked better.

At this point, the company was a group of five people backed with a $100,000 issue of convertible debentures and a sales volume, after 18 months, of $300 per month. Several major magazines were represented at the press party, but the editors did not know what to make of the company's product. The financial editor of the *New York Herald Tribune* took the plunge and featured the Loctite story in a display box in the middle of the financial page with optimistic remarks about the future significance of the new product. Thus encouraged, editors began to release articles in the trade magazines, and the flow of publicity became a flood. A box story also appeared in *Fortune.*

Initial Sales Approach

At the time, sales coverage was handled by two representatives, one in New England and one in the Cleveland area. Subsequently, the company hired a consulting firm to recommend three persons for each of 20 territories to cover the United States. Ninety days later the company had national coverage through a group of outstanding manufacturers' representatives who were to contribute significantly to the building of the company. By that time thousands of leads had come in and continued at the rate of 1000 a week. The usual response was to invite inquirers to buy a test kit. The sale of test kits was the company's principal income for the next 6 months and offset a substantial portion of expenses. The sales staff was motivated with an unusually generous commission schedule, contracted to normalize over a period of a few years. The sales approach was a demonstration to prove that the product worked and application case histories tō endorse the benefits of using it. The company, however, was overeager to prove that an anaerobic sealant could provide a lock that was strong enough to do the job and suffered by promoting products that were too strong. The new company subsequently lost some customers because of complaints from their service departments about the difficulty of taking products apart for repairs.

By early 1957, Loctite was offering a line of five different locking strengths, each in two viscosities. Complete performance data were developed along with mathematical formulas that enabled practical calculations to be made of the locking performance of any product on any threaded joint. This was the first time the designer had been offered a choice in locking torques and the chance to engineer them. It aroused a great deal of interest and an original 14-page Technical Report was reprinted verbatim in such prestigious journals as *Machine Design.*

Using a small space advertising schedule, direct mail, and, in particular, a heavy flow of case history publicity, the company barraged the metalworking industries. A stream of case history articles with suggestions for money-saving methods was produced. Industry found the ideas interesting and responded with thousands of inquiries. By and large, the company responded to these inquiries by mail because it couldn't afford to make sales calls. Only when serious interest developed on substantial potentials did sales people get involved. Thus for the first 5 years, the sales program was simply being responsive to inquiries. It was more development than sales, and it identified potentials so great the company is still far from having realized them after 20 more years of work.

Change in Approach and First Anaerobic Reformulation

A very wide range of modest-sized but very beneficial applications had now been proved. It became apparent, however, that in the aggregate, the potentials of all these proved applications were so enormously greater than present sales that a more organized approach to the market was desirable. The company simply did the obvious; that is, it selected the most promising applications and directed its sales force, which by this time was largely made up of persons making direct calls, to concentrate its efforts on these selected targets.

Emphasis on application had other consequences. It was an appropriate time to reformulate the anaerobic line using a superior new catalyst system discovered by the research chemists. And with the emphasis on application, the line was repackaged using names that identified the job for which the product was intended, such as Plastic Gasket, Pipe Sealant, Bearing Mount, Nut Lock, Stud Lock, and so forth. This made it easier for the customer to select the right product for the application it had in mind; it made it easier for distributors to sell the line; and it allowed more punch in advertising copy. Programmed sales, based on penetrating major applications, had no great impact on the company's growth, however, which continued along at 25 percent per year.

As the name Loctite became known in the metalworking industries, and company growth continued, the corporate name was officially changed to Loctite Corporation in July 1963.

European Activity

Before crossing the Ohio and Mississippi rivers, the company was active in Canada and the United Kingdom through independent distribution and shortly thereafter expanded to cover all of Europe. Loctite's associated European marketing companies began to focus on the automotive manufacturing industry, with specialists assigned to the major accounts. Marketing executives there gave top priority to the task of cultivating the relationship with the account, rather than to particular applications or to particular products. This was the period of explosive growth in the automobile industry in Europe. The sales technique proved highly successful, and this industry rose to 20 percent of Loctite's sales in four major countries. There was still low penetration of particular applications. For example, the company sold about as many cubic centimeters per car in France as in the United Kingdom. In France, it was largely for thread locking, and in the United Kingdom it was largely for sealing, though market needs were the same in both countries.

The European associates, influenced by the very successful example of the marketing company in the United Kingdom, were notably successful in their lavish use of sales people. They typically spent more than 25 percent of sales on field selling expense, raising prices to accommodate such expense, and their customers were quite willing to pay for the extra attention. In Europe the initiative continued to lie with the customer. Sales people devoted their time to solving the customers' varied assortment of fit-up problems with parts, and sales grew 40 percent per year. The unifying force behind these efforts was a barrage of advertising to explain and define the scope of the product's capabilities. In Europe, medium-sized accounts were sold direct, whereas in the United

States they were sold through distributors, and each strategy succeeded. As activity increased, the company set up European headquarters in Mijdrecht, the Netherlands, to supervise operations.

Development of Automatic Application

Throughout the worldwide automotive industry, one need of all Loctite customers was better methods of applications. A squeeze bottle in the hand is of limited use. Consequently, Loctite engineers developed automatic applicators to apply controlled doses of liquid adhesive, as small as a tenth of a drop, at rates of several thousand per hour. More specifically, they developed automatic nut coaters, automatic bolt coaters, and automatic shaft coaters. With automation the company could now prove that a 250-cubic-centimeter bottle replaced 150,000 lock nuts at a $10.00 premium, equal to a $600 per liter value.

Expansion of Marketing Goals

In recent years, Loctite has greatly expanded its marketing goals. Rather than having product managers, the company now has market managers who are concerned with organizing to best advantage the contribution that Loctite can make to each particular industry, such as automotive, appliances, small gasoline engines, steel, pulp and paper, and so forth. In the United Kingdom, where industry is so concentrated, the sales force has been divided by industry, and each market manager also manages his or her own sales force.

Second Reformulation of Anaerobic Adhesives

As of a few years ago, Loctite continued to have some nagging product problems: speed loss after shelf-aging for several years' time—an infrequent occurrence but naturally disastrous on high-speed production lines; and cure speeds that were upset by changes in plating procedures. This led to the development by a team of chemists of a second reformulation of anaerobic products in which the primer was incorporated into the adhesive, aging was eliminated, and one of the earlier limitations of anaerobics—the requirement for bond lines only a few thousandths of an inch thick—was also overcome. Futhermore, at least on threaded parts, cleaning requirements were greatly relaxed. These new products tolerate an oil film. Now it is simple for an ordinary mechanic to get good results, reliably, without technical assistance. The products thus became much easier for distributors to sell.

Dri-Loc

In the U.S. automotive industry, Loctite products were used largely to solve acute problems where nothing else would work. Automakers must have assurance that an adhesive has been applied correctly without any delay to the assembly line. They want a reliable inspection system and freedom from clean-up problems. They do not like liquids.

Loctite products weren't broadly accepted, in spite of the substantial cost reduction they offered. A team of chemists and engineers went back to the lab and came up with a solid coating of microencapsulated anaerobic liquids which could be applied to bolts at a central service center and shipped as an article of commerce. After 8 years of development, such a coating system under the name Dri-Loc first appeared. Loctite recently licensed coating plants in Detroit and Cincinnati and now the new process is beginning to get broad acceptance.

Latest Development

The newest development by Loctite chemists is a product being introduced under the Permatex label as Gel Gasket and under the Loctite label as Gasket Eliminator, which is intended to seal flanges without the use of standard die-cut gaskets. As in the case of thread locking, Loctite has been working on this application for a long time. In 1958 the first such application was on the differential cover plate of International Harvester's 24-ton trucks. There is already close to a million dollars worth of business in this field with earlier liquid gasket products. The new product has many additional features: It seals against hydrostatic pressure even before it cures; it cures rapidly in thick sections from the inside out; it releases for repairs even more easily than conventional gaskets; and the excess squeezed out of the joint stays liquid and dissolves harmlessly in the engine fluids.

Key Factors in the Development of the Loctite Line

It is important to note that the first step in commercial development was not the identification of a need but rather the development of a uniquely convenient cure that scientists thought ought to be useful "to somebody for something." "Convenience adhesives" suggest an enormous range of applications to be investigated. The art of commercial development in Loctite's case was to utilize a new technology by identifying the unique capability of the resulting product range to fill the airspace and seal the assembly. But to engineers in the metalworking industries this was a totally alien technology, and pricing to value helped finance the enormous amount of educational communication required to put across such innovative applications.

Another key step was to move rapidly into world markets in order to support the costs of continuing innovation on the broadest possible market base. Thus Loctite was selling in the United Kingdom and Canada by 1957 and by 1967 had progressed from a worldwide export agent in New York to overseas plants sourcing independent marketing companies. Today Loctite is organized as a worldwide company with all functional activities performed on each side of the Atlantic.

The company's restraint in preserving its character as a specialty company is considered to be a significant factor in Loctite's success. The so-called specialty products have the following characteristics:

1. They are low volume, differentiated products sold to a large number of customers, each buying a small amount.
2. They are products sold to performance specification for what they will do rather than on composition.
3. They are products sold under a brand name rather than a generic chemical name.
4. They are products whose cost to the user is an insignificant part of the user's total costs and far less than the savings that result from its use.
5. They are products whose applications are based on a technology that is unfamiliar to the customer, so that the user is dependent upon continued technical service.

Educational Thrust

Right from the start, the company determined to direct much of its energy and resources into educational communications, and its strategy in that area proved to be particularly successful.

Dr. Krieble had created a unique product that performed a very valuable function, but there was still the task of creating the *need* for that product. Market research initially rejected the opportunity—mechanical engineers simply did not feel the need of an adhesive to lock transmission shift lever nuts. Furthermore, there was a certain stigma attached to the very notion of using a "glue" to hold metal parts together. Marks' Handbook even now does not recognize this type of product. What Loctite offered was a chemical approach to mechanical assembly, and this was an alien technology for most mechanical engineers.

Recognizing prospective customers as a sophisticated group of engineers, technicians, and mechanics, the "high credibility" approach was chosen, rather than the "mechandising" approach of other companies. The cornerstone of this approach was the use of basic English nomenclature, rather than invented words, to name their products.

With their public relations agency, which has been with them since 1961, the company became involved in a state-of-the-art promotion. Extensive case history documentation taught engineers the benefits of Loctite technology. Stories were presented editorially in the form of feature articles in trade magazines and tabloid newspapers for easier acceptance. Up to this time, adhesives were used only as sealants and gasket dressings. Through good, reliable technical information, the company convinced engineers that anaerobics were totally engineered, predictable materials. Emphasis was placed on the ability to provide any desired predetermined constant break-loose torque to a fastener; to seal a fitting beyond the burst rating of the piping; or to apply any predetermined amount with needlelike precision through millions of cycles with complete reliability.

The aim of making the Loctite technology as "common as water" has caused the company to make even more investments in the educational process. Loctite has set up technical training and service centers in Newington and Detroit. One breakthrough in this area occurred when one of the large auto manufacturers sent 60 training managers to the Newington headquarters for instruction in the use of the company's products. To reach more remote markets, the company has converted a 40-foot moving van into a classroom on wheels. Loctite estimates that it has held 800 training sessions for 10,000 customer employees in 27 states and Mexico.

Systems Approach

The company's approach, in the OEM market at least, has been a "systems" approach to adhesive technology. It sells much more than products; it sells innovation in methods of assembly. The company often functions as a consulting laboratory for its customers. Loctite has strong competition in many parts of the world, usually based on low prices; but so great are the savings from the new Loctite assembly methods that customers are sticking with them. Customers are more interested in retaining technical services to assure successful operation of these alien chemical bonding procedures, than they are in a lower price for an adhesive product that is ony a fraction of the whole assembly cost.

In another key policy, the company has sought to promote constructive interaction between chemists and sales people. Sales staff tries new products and new applications with their customers, and chemists develop research in response to customers' needs.

Another important factor has been management's acceptance of the evolution of Loctite products into an extremely complex chemical family made up of a very large number of small uses, in which the adhesive is not only sold by the drop and even by the fraction of a drop, but where 40 or 50 kinds of drops are sold for hundreds of uses in dozens of different industries through many diverse channels.

Extensive Research and Development

The support of a large research program and the continual development of new ideas has been a major factor in Loctite's growth. Today it derives about 30 percent of its growth from products and applications that did not exist 5 years ago.

Continuing advances in anaerobic curing chemistry led to products that simplify application.

Researchers have recognized the ability of Loctite products to stretch other technologies. Anaerobic resins are being used to seal porosity in die castings and powdered metal parts, greatly expanding the capabilities of the casting industry. Loctite engineers developed vacuum impregnation equipment that it sells or rents to customers.

In improving its products, Loctite's researchers have borrowed from other technologies. In 1970, a family of thread sealants combining anaerobics with Teflon was introduced. Today, these products are a dominant factor in sealing hydraulic and other fluid systems. One major truck manufacturer uses Pipe Sealant with Teflon to seal all major fittings and recommends use of the product in service and repair.

In a recent development, adhesives that cure under ultraviolet light are emerging.

Acquisition Policy

Originally, the company thought of itself as a specialist in anaerobic chemistry, which it developed. Now it sees itself as a solver of assembly problems of parts generally, and it is this feeling that governs its acquisition policy. It will consider the acquisition of any adhesive technology that may be useful to its customers in the performance of this function.

The acquisition in 1972 of the Permatex Company, producer of the largest selling gasket compounds in the auto aftermarket, afforded Loctite the benefits of a well-established aftermarket distribution system to introduce its anaerobic gasketing material under the Permatex label. Similarly, the purchase of the Woodhill Chemical Sales Company in 1974 increased the capabilities for marketing current Loctite products. Loctite and Woodhill, the originator of an extensive line of home and car fix-it products, possess a combination of elements with mutual growth potential. Loctite conducts a strong research program. Woodhill, on the other hand, offers a broad consumer distribution system for chemically based products. Woodhill can also look to Loctite's extensive international markets for expanding its products in other countries. There are, in addition, many opportunities for cross-marketing the Woodhill automotive product lines with the Permatex distribution system. For example, Woodhill products such as epoxies and muffler-repair and rust-removing products are being added to the traditional Permatex automotive distribution system. Conversely, Woodhill will introduce a variety of Permatex products to its thousands of mass-merchandising consumer outlets.

Structures for Growth

The company operates through several major profit centers. The Americas Group, headquartered in Newington, covers the United States, Canada, Mexico and the Central and South American countries. The European Group reaches a widespread and diverse market area in the United Kingdom and the continent from its headquarters in Mijdrecht, the Netherlands. The International Group includes, Japan, Australia, New Zealand, India, and portions of Africa. The Permatex Company and the Woodhill Chemical Sales Company in Cleveland have been combined under a single management to serve the consumer and automotive aftermarket areas.

Existing facilities continue to be enlarged and updated. New distribution outlets have been set up in Spain, the Philippines, Kenya, Ghana, Nigeria, Mauritius, Southwest Africa, Mozambique, Pakistan, Cambodia, the Seychelles, and in most of the Latin American nations.

Conclusion

An overview of the Loctite Corporation now reveals an organization of 1500 employees, more than 650,000 square feet of plant and office space, and installations in 20 countries around the world. Loctite now produces and markets more than 400 prodcuts through over 75,000 trade, retail, and industrial outlets under the Loctite, Permatex and Woodhill/Duro trademarks.

Loctite's objective is to make adhesive bonding the easiest and most economical method of assembling parts. The pursuit of this goal is helping revolutionize the machine age. Not only have the products reduced costs, sped production, and increased reliability of all types of machinery, they have also permitted products and processes that were never before possible. This is perhaps one of the company's most valuable contributions.

Case History: E. I. Du Pont de Nemours & Co.–Corfam

Introduction

The Du Pont Company periodically receives requests from educational institutions, graduate students, industrial associations, and other study groups for information on various aspects of the corporate planning that went into its highly publicized "Corfam" poromeric material. The product, designed to be a replacement for leather in many uses, principally shoes, was a financial failure, and in March of 1971 the company announced that it would be discontinued. The following pages contain a brief history of the development. The venture was a relatively complex one and reasons for failure to meet the goals appear to us to be subtle and complex. For this reason, this discussion is largely historical but will attempt some conclusions and so some editorializing in an attempt to identify reasons for deviations from forecasts that may not be apparent. The reader, however, should feel free to draw from the facts any conclusions he or she deems proper; there is still considerable speculation even among those who were intimately involved with the venture about the relative importance of the conditions that contributed to its failure.

In the late 1930s and 1940s various researchers at the Du Pont Company conceived ways to make moisture-vapor permeable sheet structures.[1] Numerous uses for these structures were postulated, and among those was as replacement for leather—a material that had been traditionally associated with its ability to "breathe," interpreted by most technical people as moisture vapor permeability. Development work along different chemical lines but with the same product objective was inaugurated in two of the company's departments and proceeded with good communications and coordination between the two departments, but otherwise relatively independently, until 1955. Evaluation at that time indicated that the route being pursued by the Fabrics & Finishes Department was more apt to be commercially viable, and responsibility for further work was vested in that department. Fabrics & Finishes had held a position in coated fabrics since 1910 and in 1955 operated two coated fabrics plants; one at Newburgh, New York, producing pyroxylin and vinyl decorative coated fabrics; and one at Fairfield, Connecticut, concerned with elastomeric coated fabrics for industrial end uses. Of the Newburgh Plant's production, about 10 million square feet per year found its way into the shoe industry as findings or upper material in relatively inexpensive shoes.

Research semiworks, and pilot plant activities followed with capability to make and test material graudally improving. In October 1962, the Executive and Finance Committees of the Du Pont Company approved a project to build a commercial plant at a company site in Old Hickory, Tennessee. The plant was to have a capacity of approximately 30 million square feet per year, with reasonable expectations that refinement of the process would permit additional capacity with modest added investment. At that time, substantial quantities of material had been produced and a pilot plant, scaled to make up to 250,000 square feet per month, had been constructed at Newburgh, New York. Though sustained production at reasonable yields had not been obtained from the pilot plant, adequate quantities to test the product in commercial shoemaking plants had been produced. Goal date for production from the commercial plant was about the middle of 1964.[2]

The Early Years

"Corfam" shoes were publicly introduced at the Chicago Shoe Show in the fall of 1963. It was planned to have shoes on sale with intensive accompanying advertising, merchandising, and promotion for the spring of 1964.[3] Production of the material to support this program would come from the Newburgh pilot plant, with the expectation that the Old Hickory commercial production facility would be in operation by the summer of 1964. It had been agreed that the marketing strategy would be to introduce the material in relatively expensive shoes, to be accomplished by charging a relatively high price for the material (still considerably below the first expected cost of commercial production) and advertising those shoes that could economically justify the use of upper materials in this price class. This was designed to be fair to the shoe industry by discouraging "reaching" by a producer of shoes that could not normally justify upper materials of this price. The introduction was generally successful. Between 30 and 40 manufacturers of men's and women's shoes that normally retailed for $20.00 or more put some "Corfam" shoes in their lines.[4] Supplies were limited, because the pilot plant had never achieved efficient, full-scale production, and only a total of about 250,000 square feet of material was available, enough to make 150,000 pair of shoes.[5] Comments, primarily from shoe manufacturers and retailers, about the aesthetics of the material in relation to the expensive leathers it was replacing began to be received, but they did not appear to be particularly serious and manufacturers continued to take all of the production that was available.[6] The first material came from the Old Hickory plant in August 1964, and about 3 million square feet were delivered to shoe manufacturers in that year, at an average price of about $1.10 per square foot.

In 1965 the Old Hickory plant struggled with the process while the technical organization continually changed the product to improve it.[7] Sales reached 10 million square feet. Start-up difficulties, because of the complexity of the process, had been anticipated, and these forecasts, at least, were correct. The quality control program was adhered to rigidly, and substandard material was destroyed rather than sold. Yields were very low and costs high. Toward the end of the year, however, problem solutions began to become effective.

In 1966 about 20 million square feet were produced and sold. In this year there was considerable hope for the future of the venture. Losses decreased dramatically from those of 1965.[8] Significant progress was made on overall economics, and costs of the process were responding to the technical attention. The shoe industry showed considerable appreciation for some of the attractive properties of "Corfam."[9]

However, concurrent with efficient production of the basic "Corfam" material and attendant lower costs, the market began to demand more attractive finishes in greater variety, so that the cost of finishing increased significantly over forecast and over early estimates from the operation. Meanwhile, some concern also began to develop regarding the comfort properties of "Corfam" shoes. This is at least partially related to factors curable in the marketplace, but the steps necessary to communicate these solutions to the market, and the resistance of the marketplace to various developing aspects of the material, were not so easily overcome.

Comfort was largley related to the response of the material to deformation caused by pressure of the foot. In the case of leather, this deformation tends to be fairly permanent; "Corfam," however, is more elastic in nature. Still, "Corfam" could be fitted comfortably to the vast majority of feet, and with a well-retained attractive appearance, by careful work at the fitting stool.[10]

Another problem became evident as retailers developed resistance to the advertising and merchandising program that identified "Corfam" as a specific item of shoe upper construction. The inventory problem in shoe retailing is a serious one, with a vast number of styles and considerable fashion obsolescence. When a customer insisted not only on a particular style, but on that particular style made with a particular material, the problem increased exponentially. Superimposed over these field problems, which, in total, began to become rather important, was a fluctuating leather market that caused major increases and decreases in the drive of shoe manufacturers to use this material.

Two other influences grew. One, at least, had been in existence since the beginning of the venture, and the other appeared as a significant concern sometime in late 1966. The first had to do with the number of shoes that were imported to the United States.[11] Directly, this probably should have had no effect on the demand for "Corfam," because it was by this time being sold in Europe as well as the United States, with all production from the Old Hickory facility, but there was some inertia in the system, with a tendency for European producers of export shoes to be more reluctant to shift from leather to a synthetic material. The appeal of European shoes, principally in the case of women's shoes, had largely to do with the elegant materials and craftsmanship used in the uppers, and American buyers tended to shy away from deviation from these classical appeals. Further, marketing in Europe was at considerable economic sacrifice because of duties, fashion obsolescence, and so on, increased by the logistics.

The second had to do with the spectacular growth of vinyl coated fabrics as shoe upper materials. This plastic material, widely used for automotive and office upholstery with a leatherlike look, had found a small market (about 30 million pair per year) in the shoe industry prior to the advent of "Corfam." Indeed, Du Pont had a position with a product trademarked "Pattina" made at the Newburgh, New York, factory, which was used to replace patent leather. "Corfam" market research had carefully appraised this material by large-scale panel tests and had come to the conclusion that the comfort, flex durability, and shape retention of vinyl made it unsuitable for other than specialty shoes, such as sandals and play shoes. It was recognized that producers of vinyl-coated fabrics did have at their command techniques to color, emboss, and modify the "hand" and drape that were economical to utilize and offered infinite permutations and commutations. The advent of a poromeric material as an acceptable quality replacement for leather in shoes apparently lent respectability to the market efforts of the vinyl-coated fabrics producers. Manufacture and marketing of shoes made with this material increased spectacularly, and by the beginning of 1967 was at a rate of 100-200 million pairs per year and was still increasing exponentially.[12]

Commercial Maturity and the Beginning of Doubt

The 1967 marketing year started rather slowly. Normally material orders for fall shoes are placed during the first part of the year. Although the volume of business received exceeded that of the first half of 1966, it did not equal the forecast of expected increase. In the early summer of 1967, the marketing group made a number of special studies of demand versus price, while the production organization calculated its capability to produce, the economics of significantly increased total production, and sales at somewhat lower prices.

Throughout 1966 and 1967, various steps to increase production were made, and once process problems were resolved, these resulted in significant enlargement of the capability of the original plant. It was now estimated to have about twice the originally planned capacity. There were some relatively minor price readjustments made in the middle of 1967, principally to reposition certain types of product in the marketplace. These changes fortunately coincided with a firm leather market, and orders began to be received for very large volumes.[13]

The plant, attempting to accelerate to this production level, became involved in a monumental traffic jam, and it was not until late in the year that the forecasted quantities in the range of 4 million square feet per month were actually produced. About 24 million square feet were shipped in 1967, and at the end of the year the backlog of unshipped business totaled over 10 million square feet. However, lower prices for the product pattern in demand and excessive expense involved in gearing the plant to higher production levels resulted in losses that exceeded those of 1966.

At the beginning of 1968, the outlook for the venture was somewhat brighter. The logjam at the plant had been broken, and the backlog was high, with apparent continued interest in the product. Improved yields and operating costs gave us some cause to hope that the worst was behind us. Indeed, 1968 turned out to be the year of highest production and sales, as slightly under 35 million square feet were sold at an average price that was a little under 80 cents per square foot, and losses fell significantly. However, shortly after the spring shoe season, several adverse factors became apparent. First, the leather industry responded to the poromeric threat by producing and promoting very soft, glovelike leather that was particularly appropriate to the casual dress of the time. Leather prices softened, and the large number of imported shoes that did not use American-produced leather resulted in a great deal of price "dealing" among the U.S. tanneries.[14] Shoe industry buyers eagerly and appropriately played tannery against tannery against poromeric producer. Imports continued to increase, and vinyl shoes found a larger and larger share of the market.

Dollar sales declined almost 25 percent in 1969 and volume sales about 20 percent.[15]

Terminating the Venture

In 1970 the trend continued, and at the end of that year a careful reappraisal of the entire situation indicated the following:

1. The immediate market for poromeric material ("Corfam") in basic configurations, such as men's black, and so on, was more limited than the market research had indicated and would support only modest production, probably less than the capacity of the original Du Pont plant.
2. The market for fashion variants was highly unpredictable, and poromeric products had to compete with leathers and vinyls in which the abilities to vary "hand," tactile qualities, and other aesthetics had been developed to a more sophisticated degree than had these abilities for poromerics.
3. A substantial technical investment would be needed to commercialize Mark II, an improved product that could be varied in softness as fashion dictated, and economics of it were still uncertain.[16]
4. The business was one of high investment per annual dollar sales, like many other chemical products. It was competing in a market where the investment per annual dollar of sale was relatively small. Under these circumstances, fixed costs for the venture were high and put the poromeric offering at a competitive disadvantage in the volatile and fluctuating fashion market.
5. With production costs under reasonable control, it appeared that the venture could be made to average a small return with some losses during years of low leather prices and reasonable earnings during years of high leather prices, but that the commitment of management manpower to achieve these goals would be disproportionate to the benefits.
6. Conservative Du Pont accounting policy had depreciated the plant and established reserves in connection with other aspects of the venture, so that the extraordinary costs of terminating the operation were not very high.

This appraisal confirmed the fact that continuance of the venture would require a large amount of high-level management attention; that profits, when and if realized, would be quite modest; and that the opportunity for a poromeric product was not as great as had been originally predicted.[17]

It was decided, therefore, to terminate the operation, starting in March 1971. Negotiations had been under way with several foreign agencies, where markets were not available to us, to sell the technology, and these were continued. It seemed at first that abandonment in the United States might be an insuperable barrier to such a sale, but significantly different circumstances in eastern European countries, particularly Poland, made it feasible for that country to acquire the technology and machinery from the Old Hickory plant. Inventories were sold to an American leather brokerage firm, George Newman & Company, which has since completed negotiations to manufacture, on a relatively small scale, a poromeric product for the American market. In August of 1972, the trade name "Corfam" was sold to that company to be used on their product with a designation that would distinguish it from Du Pont's; that is, Newman was made a part of the logo over the word "Corfam." Most of the people involved with the poromeric venture, totaling about 1000 at the peak, were reassigned within the company. Some specialists who had been employed specifically because of their skills in the shoe industry found satisfactory positions with the footwear industry. A few administrative and management personnel who felt their careers might be unreasonably impeded by the time spent in the "Corfam" venture resigned to find other activities. We, of course, regretted losing these people.

Market Research—Economic Studies and Forecasting Techniques

When it first became apparent that the research program had a reasonable chance of becoming a commercial project, a modest staff of business and financial analysts, manufacturing, technical, and marketing people had been assembled as a project team to explore all the financial implications of the venture. This started about 2 years before the plant was authorized. The group was given relative freedom to utilize manpower and money necessary to produce a valid mathematical model of the expected business.

Inputs were expected to be from historical data of the shoe and leather industry, experimental data relating to our products, and judgmental factors of knowledgeable people in the shoe and chemical industries. For historical data, government and industry statistics on every envisioned aspect of shoe and leather production were collected. Experimental data were obtained by internal cost audits of the pilot plant operating capability and forecasted operating capability, and by extensive market testing, both for serviceability and perceived value, by all echelons of the shoe industry and consumers. For judgmental factors, consultants, employed from the shoe industry, as well as knowledgeable people on chemical industry economics, were utilized.

A relatively new technique known as venture analysis was used to put these data together. Mathematical modeling of this type has become more common in the intervening years, and it will be unnecessary to describe this one in detail. It was complete, attempting to bridge from the historical world to a world in which some part of the leather market would be replaced by a substitute material.[18] The model predicted that a satisfactory product that could be sold at leather-equivalent prices to the shoemaker would find an adequate market and that the initial capacity of the plant we intended to build would be economically rewarding to the company. It also indicated that capacity of considerably expanded facilities and capacity for a reasonable share of the market to be held by competitors would be in demand due to an inadequate supply of leather in years ahead (up to 20). Therefore, a satisfactory product that could be sold at leather-equivalent prices to the shoemaker would find an adequate market.

It is fitting to examine the predictions of the venture analysis model against the actual situation that existed in 1968, the fifth year of production, one normally used by the Du Pont Company for comparing the results of projects with forecast. These comparisons are:

1. The volume of material sold was approximately that originally predicted (35 versus 37 million).
2. The average price obtained for the material was approximately that predicted.
3. The cost of making the impregnated fibrous web and the cost of coating the web were approximately the predicted figures.
4. It had been found necessary to use an expensive woven interlayer in substantially all of the production, rather than an inexpensive one in about 40 percent of the production. This added between 10 and 15 percent to the total mill cost.
5. The cost of finishing, due to the great variety and the complexity and subtlety of effects needed, was from two to three times as expensive as had been predicted and added 15 to 25 percent to the mill cost of the product.
6. The technical expense of the venture consistently ran about twice the forecasted fifth year figure, and only if volume had been double that actually achieved would the cost per square foot have been as forecasted.
7. Marketing expense was about twice the forecasted figure and, again, could only have reached the expected cost per square foot at two and a half times the originally predicted fifth year volume.

It should be noted, however, that all of these interacted. A higher cost substrate might have reduced or eliminated the need for expensive woven interlayer. Greater marketing effort or more satisfactory products in the field might have made it possible to have obtained the required volume with basic colors and to have ignored the fashion market, thus lowering the finishing costs. Technical marketing expenses within our control could have been substantially lowered at the probable expense of a lower average price.

Lessons to be Learned

A number of warning signs were identified by various people who studied the problem within the company. Unfortunately, most of this occurred after the fact. There was general agreement that all of these were meaningful danger signals, but that no one of them was sufficiently dominant to be considered a "go, no go" signal in any venture.

1. When perceived quality is difficult to define and based on subjective judgment, even neutral market research findings must be examined with considerable skepticism.
2. In markets of low growth, competitive reactions will be vigorous because of the threat a newcomer offers to established business.
3. In the early years a technically complex product in a market of constant change (that is, a fashion market) will be at a disadvantage until considerable production and technical flexibility has been developed.
4. Evidence that (before commercialization) expectations of the customers are higher than the product warrants, indicates a considerable probability of some later disillusionment.
5. Failure to meet early cost and market goals, followed by failure of control procedures to correct the variance.
6. Variance from a comparable manufacturing or marketing expense of competitive products.
7. Unforecasted and major changes in market conditions.

Notes

1. "Story of Corfam," *Boot & Shoe Recorder,* October 1, 1963; "Big News in Shoes," *Reader's Digest,* March 1964.
2. Roger G. May, "Shoe Sales Are Brisk as Du Pont Launches Man-Made 'Leather'," *Wall Street Journal,* February 20, 1964; L. Lessing, "Synthetics Ride Hell Bent for Leather," *Fortune,* November 1964.
3. "Story of Corfam," May, "Shoe Sales Are Brisk," and "Big News in Shoes."
4. "Another Nylon," *Forbes,* October 15, 1964.
5. "Research–If the Shoe Fits, Another Winner for Industry," *Newsweek,* April 6, 1964; "Another Nylon."
6. May, "Shoe Sales Are Brisk."
7. E. L. Yuan, "Poromerics–How They're Manufactured," *Chemical & Engineering News,* March 9, 1970.
8. W. D. Lawson, "Status of Man-Mades," *Boot & Shoe Recorder,* June 1, 1966.
9. "Du Pont to Shift 'Corfam' Research," *Footwear News,* January 27, 1966.

10. Jackson Hand, "12 Months," *Popular Mechanics,* May 1965; "'Corfam' versus Leather for Shoe Uppers," *Consumer Reports,* November 1965.
11. "Production and Consumption of Footwear," *OECD Report 1963-1964.*
12. "Synthetics: Good Fit in Footwear," *Chemical Week,* May 1, 1965.
13. W. D. Lawson, "Status of 'Corfam'," *Boot & Shoe Recorder,* July 1967; "Stepping Lively to Keep 'Corfam' Ahead," *Chemical Week,* September 23, 1967.
14. "The Flat-Footed Shoemaker," *Forbes,*
15. "Du Pont," *Forbes,* December 15, 1969.
16. James Gubitosi, "Interview–L. S. Sinness," *Oil, Paint & Drug Reporter,* December 29, 1969.
17. Dan Cortz, "Bringing the Laboratory Down to Earth," *Fortune,* January 1971.
18. See "The Story of 'Corfam',"; "12 Months,"; W. D. Lawson, J. C. Richards, and C. A. Lynch, " 'Corfam' Research Brings Chemistry to Footwear," *Research Management,* January 1965; D. E. Abercrombie, "Venture Analysis–New Marketing Brain," *Boot & Shoe Recorder,* April 1965; "Mathematical Model Becomes Marketing Tool," *Chemical & Engineering News,* January 2, 1967; and "Despite 'Corfam'–Du Pont Is Still High on Venture Analysis," *Chemical Week,* April 21, 1972.

References

1. Boot & Shoe Recorder, October 1, 1963, "Story of 'Corfam'," Rossi.
2. Wall Street Journal, February 20, 1964, "Shoe Sales are Brisk as Du Pont Launches Man-Made 'Leather'," Roger G. May.
3. Reader's Digest, March 1964, "Big News in Shoes," Staff.
4. Newsweek, April 6, 1964, "Research–If the Shoe Fits, Another Winner for Industry," Staff.
5. Forbes, October 15, 1964, "Another Nylon," Staff.
6. Fortune, November 1964, "Synthetics Ride Hell Bent for Leather," L. Lessing.
7. Chemical Week, May 1, 1965, "Synthetics: Good Fit in Footwear," Staff.
8. Popular Mechanics, May 1965, "12 Months," Jackson Hand.
9. Consumer Reports, November 1965, " 'Corfam' Versus Leather for Shoe Uppers," Staff.
10. Footwear News, January 27, 1966, "Du Pont to Shift 'Corfam' Research," Staff.
11. Boot & Shoe Recorder, June 1, 1966, "Status of Man-Mades," W. D. Lawson.
12. Du Pont Promotion, Fall 1966, "The Fast Footrace of 'Corfam'," L. Lessing.
13. Boot & Shoe Recorder, July 1967, "Status of 'Corfam'," W. D. Lawson.
14. Chemical Week, September 23, 1967, "Stepping Lively to Keep 'Corfam' Ahead," Staff.
15. Fortune, November 1967, "Du Pont Under Pressure," Gilbert Burke.

16. Forbes, August 1, 1968, "Tattered Glory," Staff.
17. Forbes, December 15, 1969, "Du Pont," Staff.
18. Business Week, December 20, 1969, "Du Pont's 'Answer Machine'," Staff.
19. Oil, Paint & Drug Reporter, December 29, 1969, "Interview–L. S. Sinness," James Gubitosi.
20. Forbes, January 1, 1970, "Chemicals," Staff.
21. Time, January 5, 1970, "Du Pont's Troubled Dynasty," Staff.
22. Chemical & Engineering News, March 9, 1970, "Poromerics–How They're Manufactured," E. L. Yuan.
23. Forbes, "Hands Across the Sea," Staff.
24. Forbes, "The Flat Footed Shoemaker," Staff.
25. Fortune, January 1971, "Bringing the Laboratory Down to Earth," Dan Cortz.
26. Research Management, January 1965, " 'Corfam'–Fresearch Brings Chemistry to Footwear," W. D. Lawson, J. C. Richards, C. A. Lynch.
27. OECD Report, 1963-64, "Production and Consumption of Footwear," Official Publication.
28. Boot & Shoe Recorder, April 1965, "Venture Analysis–New Marketing Brain," D. E. Abercrombie.
29. Chemical and Engineering News, January 2, 1967, "Mathematical Model Becomes Marketing Tool," Staff.
30. Chemical Week, April 21, 1972, "Despite 'Corfam'–Du Pont is Still High on Venture Analysis," Staff.
31. Du Pont Magazine, September-October 1966, "New Fashions Afoot," Staff.
32. Time, November 27, 1964, "The Master Technicians," Staff.
33. University of Illinois, " 'Corfam', An Advertising Case History," Prof. James E. Moyer.
34. Time, April 3, 1964, "The Synthetic Shoe-In," Staff.

Case History: PPG Industries–Silica Pigments

Silica Pigments

The pioneering development of fine particle precipitated silicas and silicates by PPG Industries, Inc., Chemical Division, has spanned over 30 years. In retrospect, this development exhibits the essentials now defined for the initiation and pursuit of a successful commercial chemical development, albeit in slow and labored form.

The experiences of the R&D and marketing personnel of PPG at the original plant of the Chemical Division in Barberton, Ohio, in marketing a precipitated calcium carbonate rubber filler, Calcene®, to the adjacent rubber industry in Akron in the early 1930s provided seminal insights. New product needs and market opportunities, as well as economic prospects for the new development, became apparent.

Thus, technical insights called for (1) a white pigmentary material finer than Calcene, and more nearly comparable in particle size to carbon black, to provide greater rubber reinforcement; such a product could (2) satisfy the need for greater color flexibility in many rubber products then limited to black due to the use of carbon black reinforcement for adequate strength and durability. An exploration of other low-cost chemicals to replace the Solvay soda ash precipitant, which had been employed with by-product calcium chloride from the soda ash plant to produce Calcene, was undertaken in the hope of achieving these objectives thus conceived.

The development of not just one product, but a series of products, with stepwise improvements has resulted. Close coordination was established early among R&D, production, and marketing personnel and has been continued and strengthened during this development to support and miantain favorable progress. The family of products now marketed serves needs especially in the rubber, paper, and coatings industries. Termed Silenes® and Hi-Sils®, these fine particle materials function as reinforcing fillers or "pigments" in rubber, as true pigments and pigment extenders in paper, and as pigment extenders or flatting agents in coatings. Other areas of application are under exploration. The specific product developments are outlined below.

Silene and Hi-Sil Pigments (Fillers) for the Rubber Industry

Suffice it to say, the early R&D effort was successful in 1937 in demonstrating that the replacement of soda ash (that is, sodium carbonate, solution by sodium silicate solution) as precipitant in the reaction with calcium chloride under selected conditions gave a hydrated calcium silicate, $CaO(SiO_2)x$, of essentially spherical ultimate particles about 100 millimicrons in diameter; this product was considerably finer than Calcene, although still coarser than carbon black. After washing, drying, and milling, the white powder reinforced rubber to a level not possible with Calcene. Happily, the sodium silicate could be made from a deposit of good-quality silica sand on hand by reaction with soda ash; this would be done in a glass-making type furnace using know-how already possessed by PPG. New process technology was developed by the PPG engineers for producing the new product, which was first made in the Calcene pilot plant. After promising consumer trials, the pigment was marketed from a commercial plant as Silene in 1939.

Thus, PPG was able to utilize low-cost raw materials to provide an improved product with desired performance properties on a substantial scale and to keep the price low and competitive with the appropriate grades of carbon black.

The weaker synthetic rubber available during World War II, after the supply of natural rubber was cut off, forced manufacturers to reconsider their raw materials. Users quickly confirmed the improved reinforcement obtainable in light and colored rubber articles by using Silene. The Navy complained that shoe soles loaded with carbon black were marking up ship decks; housewives also complained about black marks on their floors resulting from the soles and heels being made with natural rubber and carbon black. To overcome this problem, a large rubber company combined a new synthetic elastomer with Silene to make a leatherlike soling that did not mark floors and provided improved wear resistance.

PPG at first possessed minimal rubber laboratory applications facilities for testing of Calcene, but in 1935 the company added further equipment and hired two men from rubber fabricators to staff a new laboratory. Subsequently, one of these men, knowledgeable in contacts with the rubber industry, stepped out of the laboratory, along with one of his research-chemist companions, to form the first commercial development team for Silene. A third companion on this team was a member of the marketing group in the Pittsburgh General Office of PPG. However, because the company's line sales force was not familiar with the new product, nor with the market, PPG turned the sales of Silene over to a rubber-industry-oriented Akron distributor. Later, PPG trained its own technical service and sales personnel.

The close coordination between the commercial development group and the R&D personnel stimulated continuing research on the Silene process; this resulted in an improved Silene of finer ultimate particle size, averaging approximately 30 millimicrons in diameter. This improved product was introduced commercially in 1944. It replaced the earlier Silene rather rapidly due to the improved reinforcement that resulted and provided significantly improved strength in light-colored articles of the large-scale synthetic rubbers, especially GR-S (now termed SBR).

After World War II, furnace blacks with better reinforcing properties than the earlier channel blacks were introduced. PPG R&D personnel, responding to the need for a still finer and more highly reinforcing pigment than Silene EF, were able to modify the Silene process, removing most of the calcium with hydrochloric acid, and produced a fine, amorphous, hydrated silica. The new product was less agglomerated and easier to disperse than Silene. It was first produced in the Silene plant and was introduced as Hi-Sil in 1948. The product was later known as Hi-Sil 101.

Hi-Sil 101 was succeeded by products produced by direct acidification of sodium silicate without intermediate precipitation of calcium silicate. Carefully controlled conditions were developed for close control of important properties, such as surface area and particle size, and low-cost carbon dioxide was used as acidification agent. Thus, Hi-Sil 101 was succeeded by Hi-Sil C, also designated Hi-Sil 202, and finally in 1955 by the improved product, Hi-Sil 233. The white, powdered Hi-Sil 233, the current product, possesses an ultimate particle size of 22 millimicrons, comparable to that of the finest furnace blacks.

The product is also supplied in compressed, deaerated bags or as pellets. These Hi-Sils were the first reinforcing hydrated silicas available at prices competitive with the established carbon blacks. Production was supplied from a new expanded plant constructed in 1954. In 1969, PPG built a second silica pigments plant at Lake Charles, Louisiana. One experimental ultra-pure hydrated silica, Hi-Sil X303, was partially developed in 1953 and 1954 for use in reinforcement of silicone rubber but was withdrawn due to unfavorable factors.

PPG continued to invest R&D funds in development of these fine particle materials because of the promise of increased sales. The R&D staff working on the silicate pigments numbered about a half dozen in the late 1930s. During the development of the Silene pigment, the pilot plant was operated by about a dozen men in round-the-clock shifts for approximately 2 years. The R&D staff was gradually enlarged during the development of the Hi-Sils to a peak of 35 people by the early 1960s, and the commercial development and technical service personnel contacting the rubber industry increased to six people.

The rubber laboratory was further enlarged with additional mixing, shaping, curing, and testing equipment to the point that it surpassed the facilities of all but the large rubber companies. The pigment pilot plant became one of the best pilot plants in the country; it was equipped with its own electron microscope for product observation and control.

PPG, Chemical Division, was basically a chlorine-alkali producer, marketing primarily large-volume, bulk commodity chemicals. Some reorientation was required in the development of the performance-type silica and silicate pigments; however, efforts were steered toward making a few large-volume pigment products. Processing problems in washing, drying, and milling of the products were overcome in ways to maintain low costs and thus low and competitive prices.

During the development of the Silene and Hi-Sil products, the commercial development and rubber compounding laboratory groups worked closely together. Being knowledgeable, they suggested the more likely applications, often influenced by responses from rubber industry development personnel communicating their needs. PPG chemists published trade-journal articles comparing the properties of the PPG pigmentary fillers with the properties of other reinforcing fine particle materials familiar to the rubber industry in typical formulations. Numerous seminars were presented to rubber industry groups to assist in communicating this information.

Industry use of the PPG Hi-Sils occurred first in applications offering significantly superior properties. An important advance was in partial replacement of carbon black with Hi-Sil 233 in the off-the-road tires in 1956 to obtain reduced chipping and improved tear strength. Combinations of the Hi-Sils with carbon black were useful with only minor modification of the compounding additives. When used alone as the reinforcing filler or pigment, the Hi-Sils required changed compounding formulations as compared with carbon black. These were investigated extensively by the PPG chemists, who provided recommended formulations to rubber industry technical personnel. The different curing behavior of the Hi-Sils is ascribed to the selective adsorption or tendency for surface reaction of some of the conventional compounding additives in ways that are different from their behaivor with carbon blacks.

Compounding studies were carried out using Hi-Sil alone and mixed with carbon black, with production and road-testing of experimental tires. The road tests showed average tread-wear resistance readings ranging from 60 to 85 percent of control treads made with carbon black, when Hi-Sil was used as the filler alone. When Hi-Sil/carbon black mixtures were used, the performance ratings were higher. Road-wear performance equal to carbon black was finally obtained with Hi-Sil 233 alone as reinforcing pigment using a mercaptosilane coupling agent. Although not currently economical for use in tire treads, the behavior with this coupling agent has encouraged replacement of carbon black in some uses and suggests a direction for future research in an endeavor to enlarge the use of Hi-Sil in tire applications.

Currently, the fine particle silica pigment is finding increasing use in commercial tires to promote higher adhesion to tire cord. Other major applications are in producing quality shoe heels and soles, wire and cable compounds, and mechanical goods.

The commercialization of the Hi-Sil pigments required considerable expenditure for toxicological tests. These tests, conducted at independent laboratories, showed that PPG's hydrated silica did not pose the silicosis threat characteristic of crystalline forms of silica.

During pursuit of the development of PPG's fine silicas, technical information exchange agreements with large European companies who had entered the field and were developing similar products were of considerable assistance. The combined technical resources thus represented were helpful in solving some of the more difficult problems and in considering new pigment modifications for new applications.

PPG recognized early the importance of patents in protecting the new development and was able to obtain good patent coverage of the new products and processes. Although most of these patents have now expired, the specialized know-how in the production, processing, and application of these materials represents a substantial resource.

More recently, beginning in 1967, PPG developed Hi-Sil EP pigment to expand the selection of the pigmentary silica products offered to the rubber industry. This product possesses a larger particle size than Hi-Sil 233, but offers easier processing and certain other advantages. The suitability of this product in various applications has been tested extensively, with broad sampling of the rubber industry fabrication groups by PPG's market development personnel.

The product has not yet (early 1974) been produced on a continuous commercial scale. However, the shortage of energy and carbon black are expected to speed development of this new product toward a fully commercial status and to increase the volume of use of Hi-Sil 233 as well.

In summary, it can be said that the commercial development of the Silenes and Hi-Sils by PPG has embraced many of the favorable features regarded as essential for successful commercial chemical development. Although the development period has been long, the negative aspects probably must be accepted in the introduction of new raw materials into an industry producing high performance products. The substantial technical backup in such a development must be justified by the ultimate opportunity that appears to be offered.

Hi-Sil Pigment and Pigment Extender for the Paper Industry

As an established supplier of chlorine, alkalies, and other large-volume processing chemicals to the paper industry, it was opportune for PPG to encourage testing of its rubber pigments, the Silenes and Hi-Sils, as alternatives to the clays, calcium carbonates, and other fine particle materials used in paper manufacture. When one paper company repeated tonnage orders for Silene, PPG employed an experienced paper chemist to investigate and expand the application in this industry.

His laboratory tests showed that the calcium silicate pigment contributed opacity and brightness to paper and that these effects were substantial, especially when the pigment was used as an "extender" for titanium dioxide pigment. Experimental plant trials with Silene pigment combined with titanium dioxide in the plant of a large magazine publisher were encouraging; results indicated that the weight of the paper sheet might be reduced without sacrificing optical properties or printability and that the pigment cost, as well as total paper cost, would be reduced.

Most paper companies were at first only mildly interested in the Silene pigment, doubting its value because of the lower index of refraction than titanium dioxide. However, PPG chemists demonstrated the benefits of the Silene pigment in seminar presentations, using light-scattering theory and procedures to show the improved opacification and brightness; the latter resulted from an increase in light-scattering interfaces when the pigment was used alone in making paper, and especially when used with titanium dioxide.

Subsequently, research was conducted to determine the characteristics of a silica pigment that would exhibit optical and other properties best suited for use in paper. By increasing the particle size to obtain maximum light-scattering and by reducing the calcium content, an efficient new pigment for use in paper, Hi-Sil 404, was developed.

Again, as in development of the new rubber pigments, the paper laboratory facilities were expanded and three additional personnel employed for a concentrated 2-year program of paper-pigment testing. A team of four commercial development and technical service engineers with paper industry experience encouraged and assisted mill trials, suggesting the types of paper and paperboard that would benefit most from the properties of the new pigment. Success in the trials in small mills, often with one specialty grade of paper, then encouraged trials of the new pigment in other paper grades.

New pigment technology was developed to produce Hi-Sil 404. The pigment was first produced in experimental quantities in pilot plant facilities adapted from those used in the rubber pigment development. The product was promoted from a development to a line product in 1963 and was then supplied from commercial facilities.

PPG engineers responded to needs in paper-mill use of the new pigment by contributing special handling procedures; for example, the pigment was packaged in white paper "beater" bags that could be added unopened directly to the paper-making process. For large-volume consumers, PPG developed a slurry-unloading technique, using water from the paper mill to unload the powdered pigment from a hopper car.

Competition developed when a competitive pigment manufacturer, observing PPG's progress with the new material, offered a similar sodium aluminosilicate product at a slightly lower price. Later on, other pigments such as hydrated aluminas entered the competition for this market. PPG met these challenges, however, while increasing sales volume. Due to its high optical efficiency, Hi-Sil 404 pigment has maintained its position as the major extender-type pigment used by the paper industry.

Hi-Sil Pigment Extender and Flatting Agents for the Paint Industry

During the early 1960s, PPG explored the use of its silica pigments in paints, utilizing laboratory facilities serving the development of its own process for titanium dioxide pigment. Application as extenders for titanium dioxide, which could improve opacity, hiding power, and consistency and lower costs were sought.

As was anticipated, the scientific base of information relating particle size to optical properties, which served in the development of the efficient pigmentary Hi-Sil 404 for use in paper, was also found useful in selecting an improved paint pigment extender. Hi-Sil 404 proved to be better than Hi-Sil 233 and other experimental pigment variations tested. By adding a step to further purify the product, Hi-Sil 422 pigment was created for the paint industry. Modification of the Hi-Sil 404 process required only one man-year of laboratory work in establishing the new pigment; also, very little additional pilot plant work was needed. A premium price for this pigment was justified by the extra purification step.

In the market development, PPG's research chemists had to provide convincing demonstrations, with data, to the commercial development group showing that the new Hi-Sil pigment had a potential in paints. The first commercial development contact was with research personnel at PPG's Coatings and Resins Division. Chemical Division research personnel presented a seminar on the performance of the proposed new pigment, with demonstration paint panels. The efficiency of the new pigment was greater in latex than in oil-base paints. The first commercial application that followed was by the Coatings and Resins Division in latex wall paint. This was, fortunately for the new silica, an expanding market.

A full-time commercial development person was subsequently appointed for Hi-Sil 422; he was armed with technical bulletins containing suggested paint formulations developed by the research group. The latter group also prepared articles for trade publications and presented talks to paint society groups. Several PPG salesmen being trained to sell titanium dioxide sharpened their wits between 1965 and 1968 by selling Hi-Sil 422 pigment. Later, the shortage of titanium dioxide with its rapid price climb starting in 1972 has given a substantial boost to the sale of Hi-Sil 422 pigment. Through Hi-Sil 422 PPG shares the titanium dioxide extender market with modified silicas, sodium aluminosilicate, and fine particle clays.

PPG also built on an initial small application of Hi-Sil 233 pigment as a flatting agent in paints, having been encouraged in this direction by the fact that precipitated silicas are efficient gloss reducers and disperse easily in all types of paints.

However, a range of products is required because of the wide variety of coatings using flatting agents; the challenge to Chemical Division R&D was to obtain the right particle size distribution in each of several silicas to cover the desired range. Experimental work extended over a period of about 5 years. Three silica flatting agents differing with respect to particle size distribution have been made on a pilot plant scale. The processes involve the use of additional equipment at the end of previously established Hi-Sil pigment manufacture to modify a small portion of the product. Three new flatting agent products designated Lo-Vel TM 27, 29, and 39A are now in full commercial production. They are at present being purchased by several customers, including PPG's Coatings and Resins Division.

As in the development of the silicate and silica pigments for rubber, PPG has maintained close coordination among the contributing company groups—R&D, production, market development, patents, and sales—in the development and adaptation of these new products to serve needs in the paper and paint industries.

Case History: General Electric Company—Noryl Resins

Introduction

Noryl® resins are the results of a relatively new chemistry of oxidative coupling invented by Dr. Allan S. Hay of the General Electric Corporate R&D in 1957. The resins are based on polyphenylene oxide (PPO®)—the product of oxidative coupling of 2,6-xylenol.

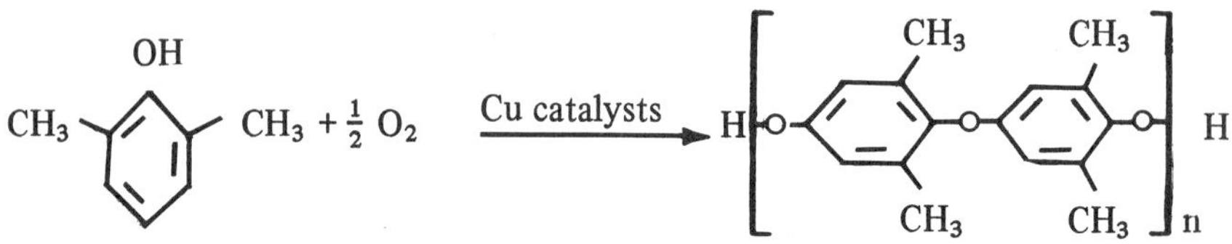

2,6-xylenol is produced by catalytic alkylation of phenol with methanol in a fixed bed reactor.

The resins and the monomer are produced by General Electric, whereas the basic processes were developed in the early and mid 1960s. Numerous major process improvements and modifications have been implemented since, with a substantial enhancement of product variety, quality, properties, and economics.

Chronology

1956 PPO® polymer discovered by Dr. Allan S. Hay, of the Gerneral Electric Research Laboratory, Schenectady, New York.

1958 First chemical engineering scale-up at Research Laboratory.

1960 Project responsibility transferred to Chemical Development Operation, Pittsfield, Massachusetts.

1961 First miniplant (10,000 pounds per year) PPO polymer production in Pittsfield. Process development work begins.

1961 Research Laboratory begins synthetic 2,6-xylenol process studies.

1962 Extrusion and molding studies begin. Polymer physical characteristics take top priority.

1963 Polymer Products Operation formed within Chemical Development Operation to commercialize products of oxidative coupling.

1963 Decision made to convert former polycarbonate semiworks facility for production of PPO polymer.

1963 One million pounds per year of crude 2,6-xylenol found in England.

1963 Decision made to construct several-million-pounds-per-year purification plant using new General Electric technology.

1963 Sixty percent-owned joint vernture established with Algemene-Kunstizijde Unie (AKU) in Holland. Company called NV Polychemie AKU–GE

1964 Polymer and monomer purification semiworks come on stream.

1964 Decision made to build synthetic 2,6-xylenol reaction and recycle system to match purification plant size.

1964 PPO polymer commercially introduced in New York City in December.

1965 Decision made to build a multimillion-pound-per-year commercial PPO polymer plant in Selkirk, New York, on newly acquired 500-acre site.

1965 Monomer synthesis plant comes on stream. General Electric now integrated back to phenol with a proprietary process.

1966 Introduction of Noryl resins, major product of the Polymer Products Operation, occurs simultaneously in June in New York City and Arnhem, Holland, and Tokyo.

1966 Commercial PPO-Noryl resins plant comes on stream.

1970 A new Noryl plant constructed in Bergen op Zoom, Holland.

1971 A joint venture with Nagase Co. in Japan established to manufacture and sell Noryl resins in Japan.

1975 A major expansion completed in Selkirk, New York.

Polymer Development

Following Dr. Allan S. Hay's successful bench-scale reactions, the first chemical engineering work on PPO® polymer was conducted at the General Electric Research Laboratory in 1957 and 1958. The process at that time was used exclusively to produce sufficient amounts of polymer for evaluation. No attempt had been made to optimize the process, and it therefore consisted of several cumbersome steps and used expensive solvents and catalysts.

Between 1958 and 1960, a number of laboratory tests at the Research Laboratory and a small amount of chemical engineering at General Electric's Chemical Development Operation (then a component of the Chemical and Metallurgical Division) refined the process to the point where more reasonable ingredients could be used. A six-step process was available for the first miniplant runs.

From 1960 to 1962, a chemical engineering effort in the Chemical Development Operation was concentrated on optimizing each of the above steps in a 10,000-pound-per-year multipurpose miniplant. The aim during this period was twofold: (1) process optimization and (2) sufficient polymer of uniform quality. These steps were required to characterize the material, so that a business decision on the next step—the scale-up to the million-pound-per-year semiworks plant—could be justified not only from a process but also from a product quality standpoint.

Much of the equipment development took place in the 1-million-pound-per-year semiworks between January 1964 and April 1965, when the final design was frozen for the commercial plant. By this time, the chemical engineers had reduced the initial process from a six-step system to a streamlined three or four-step operation. The commercial plant was designed and constructed by General Electric's Polymer Products Operation* engineers in conjunction with Blaw-Knox. This grass roots facility was designed and constructed in 16 months on a 500-acre site in Selkirk, New York. This "first of a kind" plant, capable of producing both PPO and Noryl resins, operated at the design capacity from the outset.

Monomer Development

Although General Electric had developed a process for the oxidation of 2,6-xylemol, this monomer was not available in any substantial quantity with the desired purity. The early laboratory work in Schenectady from 1956 to 1960 and the first Pittsfield miniplant work in 1960 and 1961 used monomer supplied from a specialty chemical house at $2.50 per pound. This 2,6-xylenol was obtained by multiple crystallization of crude coal tars, and not more than 10,000-20,000 pounds per year could be made available by the specialty firm. As the product properties looked more and more favorable and a fit in the marketplace was identified, three particular problems faced General Electric's Chemical and Metallurgical Division:

1. An economic route to synthesizing 2,6-xylenol had to be developed because none existed.
2. A source of at least crude 2,6-xylenol had to be available for development production of PPO.
3. A purification process to provide the ultra-high-purity, polymerization grade monomer had to be developed, first, to purify purchased crude and eventually to be integrated into the synthesis process.

The project was essentially divided between teams at the Research Laboratory and the Chemical Development Operation. The Research Laboratory concentrated on uncovering the synthetic route to 2,6-xylenol, while Chemical Development Operation engineers concentrated on finding a crude coal tar source and a purification scheme to produce polymer grade xylenol.

Ordinarily, a project of this type is usually focused first on synthesis and subsequently on purification. In this case, however, the immediate need for a high-purity monomer to produce the polymer—coupled with the identification

*Noryl resins are produced today by the Noryl Products Section of General Electric's Plastics Business Division.

of a crude source—required initial concentration on purification while additional development work was conducted in parallel on synthesis.

Between 1962 and 1964, worldwide searches for coal tars containing 2,6-xylenol found in England approximately 1 million pounds of crude material in the 59-90 percent purity range. During this same period, the Chemical Development Operation worked on a variety of purification schemes. Distillation was tried first but was quickly ruled out because of the very narrow spread in relative volatilities between the various phenolic components. In fact, the meta- and paracresol fractions have almost the same boiling point as 2,6-xylenol over the pressure range available.

After rejecting distillation, the engineers turned to liquid-liquid extraction. It was known that aqueous caustic treatment of alkylated phenols result in selective distribution of the various phenols. The Chemical Development Operation's intermittent work on this project from August 1961 through early 1963 developed a process that clearly identified liquid-liquid extraction as the favored route. Here, the partition coefficients were most favorable and obtaining ultra-high-purity 2,6-xylenol (less than ten parts impurities per million) by this unit operation looked feasible.

A continuous laboratory extraction system was installed early in 1963, and the decision was made to design and build a liquid-liquid extraction pilot plant with a capacity of several million pounds in Mount Vernon, Indiana. The design and construction of this plant was done solely by Chemical Development Operation engineers, and this "first of a kind" plant came on stream with no difficulty in May 1964. This design was a several hundredfold scale-up from glassware extraction equipment into a several-million-pound plant.

Thus, with 1 million pounds of crude available, several million pounds of purification capacity in place, and a 1-million-pound polymer semiworks in place, the knot was tied, and sufficient material was available for the commercial announcement of PPO polymer in December 1964. The remaining issue was to make sufficient amounts of monomer available. This called for implementing the synthesis know-how that had been accumulating in parallel with the purification development.

Most alkylations of phenol reported in the literature were vapor phase reactions of methanol with phenol over fixed-bed catalyst producing a variety of alkylated products, including ortho-, meta-, and paracresol, and various xylenol isomers. Although these processes would have been satisfactory for cresylic acid producers who sell a variety of mixed phenols, the General Electric process had to be extremely selective toward 2,6-xylenol formation. No by-product credits could be taken, and the finished product economics dictated that a monomer had to be found for under 20 cents per pound. Several routes to the process were taken, and after numerous novel syntheses were investigated, two processes were settled upon for extensive study—the alkylation of phenol with methanol and the reaction of cyclohexanone with formaldehyde.

The final process selection was based on a breakthrough at the General Electric Corporate R&D Center.

The principal contribution of the Research Center was the finding that a specific basic catalyst was extremely active and selective when temperature was controlled over a very narrow range. The tight temperature control required for high activity/selectivity made the reactor design somewhat critical. It was finally decided to build a multitubed, fixed-bed reactor with temperature control by a molten salt system. This particular reaction system gave extremely good phenol yields but resulted in a poor utilization of methanol because of the high temperature required for activity. Nevertheless, the need for a greater source of 2,6-xylenol required taking the next step of building a several-million-pound-per-year semiworks to complement the existing purification system in Mount Vernon, Indiana. This plant, designed by The Lummus Company, came on stream in the summer of 1965. Although the process gave phenol yields of over 85 percent to 2,6-xylenol, the methanol yields never reached 50 percent. Parallel laboratory studies resulted in defining a moderate pressure reaction (100 pounds per square inch), which kept the phenol selectivity high, while lowering the reaction temperature 30°-50°C. This decreased the rate of the methanol decomposition reaction and substantially improved the methanol and phenol selectivities.

Thus, a process was defined that would be able to provide 2,6-xylenol for well under the original goal of 20 cents per pound.

In summary, the chemical engineers had to develop three fundamentally new processes from new chemistry and build "first of a kind" plants. These processes involve every unit operation conceivable. All processes are now considered to be economically attractive, plants have undergone several expansions and in each instance are operating at or above design capacity. At no time in the development of this total business picture has the chemical engineer been the roadblock to its success. On the contrary, the engineer has taken the lead in bringing the project to the favorable position it enjoys today.

Product Planning

The first product to result from General Electric's oxidative coupling technology was PPO polymer, an engineering thermoplastic that was announced in December 1964. PPO had a property profile that was distinctly different from other commercial engineering plastics materials, such as the polyacetals (Du Pont's DELRIN and Celanese's CELCON); the nylon molding compounds; Union Carbide's POLYSULFONE; and the polycarbonates (General Electric's LEXAN and Mobay's MERLON).

PPO polymer extended the property profile available in existing engineering thermoplastics in several ways. For example, it has a brittle point of −275°F and a heat distortion temperature of 375°F. This temperature use range was about a hundred degrees Fahrenheit greater at each end of the temperature spectrum than that offered by other engineering plastics.

In addition, the excellent hydrolytic stability of PPO polymer at elevated temperatures indicated that this new material might be useful in hot-water transmission, both industrially and in the home, a field presently dominated by metal.

Another property of PPO polymer—its extremely low creep—promised significant industrial uses. For example, at 3000 pounds per square inch after 300 hours, its creep was still well under 1 percent.

Finally, the excellent electrical characteristics of PPO polymer—particularly its unexcelled low loss over a wide frequency and temperature range—marked it as an excellent candidate for a variety of electrical applications heretofore reserved for the much more expensive fluorocarbons.

One important consideration was that General Electric already had achieved a significant position in the engineering plastics business with its LEXAN polycarbonate resins. From the very beginning of the PPO polymer project, the product planners had recognized that there would be some overlap between the two materials. However, it had also been correctly anticipated that the real property differences would lead to distinct and significant opportunities for each. The transparency and unmatched toughness of LEXAN had been responsible for its rapid acceptance in a variety of appliance, safety, and automotive applications. Similarly, the unique combination of thermal, electrical, fire-resistant, and hydrolytic properties indicated that PPO-based polymers would find their own application niche with only modest overlap with other engineering materials.

However, it was recognized that the fear of overlap could act as an inhibiting influence in the marketing development effort. It was partly for that reason—as well as to provide direct control and unidivided responsibility in all aspects of the project—that the Polymer Products Operation had been established within the Chemical Development Operation. The specialized PPO market development team focused their efforts in three application areas, electrical-electronic, aqueous transmission, and structural components. Their work was highly successful, and a number of commercial applications were developed.

During the course of that marketing program it became apparent that a far larger and broader opportunity was available to a PPO related material that could fill the significant price/property void between such commodity plastics as high-impact styrene and ABS and the other available engineering plastics. In late 1965, the Polymer Products Marketing personnel defined the target as a product with a heat distortion of 250°-275°F, good hydrolytic stability and flame resistance, and a selling price of 2-3 cents per cubic inch.

The flexibility of oxidative coupling chemistry allowed the General Electric researchers to develop the modified product—NORYL thermoplastic resin—in less than 4 months. By the second quarter of 1966, changes were made in the Selkirk, New York, plant to incorporate NORYL resins into their manufacturing operations. In June 1966, the new product was introduced simultaneously at the National Plastics Exposition in New York City, in Europe, and in Japan.

NORYL resin combines an overall balance of properties with excellent processability. It is a tough, rigid material that maintains excellent mechanical and electrical properties over a wide temperature range. It also exhibits good dimensional stability with low creep and low moisture absorption. It is self-extinguishing; a new grade meets UL group I requirements.

The new product found wide acceptance in the area of cold- and warm-water applications including showerheads, water sprinklers, and pump components. Within the electrical and electronics industries, NORYL resin was used for business machine housings and television components, including deflection yoke parts, tube sockets, tuner strips, high voltage cups, and connectors.

As predicted by the marketing specialists, NORYL resin has filled an existing gap and has offered the design engineer a new degree of flexibility. NORYL resin had immediate worldwide market acceptance, and sales have since grown at a very rapid rate.

Since the late 1960s, a key to the growth of the NORYL resins has been consumer demand for improved product safety in consumer electrical and electronic applications. For these uses, new NORYL formulations were developed that possess a balance of flame retardancy and electrical characteristics unmatched by any other resin.

Meanwhile, the impact strength of NORYL has been significantly upgraded. Whereas the original NORYL resins had impact strengths of 2.5 foot-pounds, they now range up to 5 foot-pounds—well above most competing materials and useful in many high-abuse applications.

The basic properties of the resin, along with the improvements made in the original products (fire retardancy, impact resistance, and processability), have led to an even larger and broader market for NORYL than originally visualized. The major application areas are in small and large applicances, communications, business machines, automobiles, and consumer electronics. New foamable grades are finding use in large housings and sound-deadening applications; and recently introduced extrusion resins are headed for construction, processing, and other industrial uses.

Equally important, NORYL is worldwide in scope. The major manufacturing location is at Selkirk, New York, which is the business headquarters. There are additional manufacturing facilities in Holland, in Japan (joint venture with Nagase), and in Australia. The product is sold throughout the world and continues to gain acceptance in a growing number of high-performance uses.

The NORYL business today is part of General Electric's Plastics' Division, which supplies the broadest line of engineering plastics in the industry, including LEXAN polycarbonate resins, LEXAN sheet, VALOX polyester resin and GENAL phenolic molding compound.

Case Histories Presented at CDA Seminars

Robert D. Bent, ARCO Chemical Company. "Future Trends." Opportunity and Risk in Commercial Development, 33d Annual Meeting (CDA), March 1977.

Edward Donley, President, Air Products and Chemicals, Inc. "Commercial Development at Air Products and Chemicals, Inc." CDA, March 1971.

K. C. Esty, Mobil Oil Corporation. "Marketing Tomorrow's Automotive Oil Today." CDA, March 1977.

P. T. Foss, Midwest Regional Director, Eastman Kodak Company. "The Pocket Instamatic Camera Story."

L. F. Gayner, Manager, Home and Personal Care Products, The Dow Chemical Company. "Caustic Soda or Saran Wrap.®" CDA, March, 1975.

R. J. Hughes, Union Carbide Corporation. "Opportunities and Risks in Commercial Development." CDA, March, 1977.

John O. Logan, UOP, Inc. "Develop or Acquire?" Opportunity and Risk in Commercial Development, 33d Annual Meeting (CDA), March 1977.

E. M. Milner, Director Products Technology, Astroturf, Monsanto Co. "The Astroturf Story." CDA, November 1972.

Clarence M. Neher, Ethyl Corporation "Opportunities and Risks." Opportunity and Risk in Commercial Development, 33d Annual Meeting (CDA), March 1977.

Charles E. Reed, General Electric Company. "Strategic Planning in an Uncertain Environment." Opportunity and Risk in Commercial Development, 33d Annual Meeting (CDA), March 1977.

Robert D. Rosenthal. Chairman of the Board and President, Neo-Tec Corporation. "The Transition of an Aerospace Company into a Leader in Chemical Processing Instrumentation." CDA, December 1974.

R. D. Schneider, Stauffer Chemical Company, Inc. "Risks and Opportunities Encountered During the Formation and Growth of the Food Ingredients Div. of Stauffer Chemical Company." CDA, March 1977.

P. J. Webster, Vice President, Corporate Communications, Damon Corporation, "Damon Corporation: A Diversified Healthcare Company with an Outstanding Growth Record." CDA, March 1973.

BIBLIOGRAPHY

Books

Alberts, William W. and Segall, Joel E. eds. *The Corporate Merger.* Chicago: The University of Chicago Press, 1966.

Allison, David, ed. *The R&D Game.* Cambridge, Mass.: M.I.T. Press, 1969.

Baker, Michael J. *Marketing New Industrial Products.* London: MacMillan, 1975.

Blood, Jerome W. *Utilizing R&D By-Products.* New York: American Management Association, 1967.

Bright, James R. *Research, Development, and Technological Innovation,* Homewood, Ill.: Richard D. Irwin, 1964.

Buell, Victor P., ed. *Handbook of Modern Marketing.* New York: McGraw-Hill, 1970.

Chacko, George K. *Technological Forecontrol.* New York: American Elsevier, 1975.

The Conditions For Success in Technological Innovation. Paris: Organization for Economic Co-Operation and Development, 1971.

Coram, T. C., and Hill, R. W., eds. *New Ideas in Industrial Marketing.* London: Staples Press, 1970.

Corey, E. Raymond. *Industrial Marketing: Cases and Concepts,* 2nd ed. Englewood Cliffs, N.J.: Prentice-Hall, 1976.

Drayton, Clarence I., Jr., et al., under the direction of G. Richard Young. *Mergers and Acquisitions: Planning and Action.* New York, Financial Executives Research Foundation, Inc., 1963.

Fishlock, David. *The Business of Science.* New York: John Wiley, 1975.

Gabor, Dennis. *Innovations: Scientific, Technological and Social.* London: Oxford University Press, 1970.

Gerstenfeld, Arthur. *Effective Managementing of Reseaarch and Development.* Reading, Mass.: Addison-Wesley, 1970.

Giragosian, N. H. *Marketing Guide to the Chemical Industry.* Technomic Publishing Company, Westport, Conn., 1965.

Grabowski, Henry George. *The Determinants and Effects of Industrial Research and Development.* Ph.D. Thesis, Princeton University, 1966. Ann Arbor, Mich.: University Microfilms, 1967.

Harvey, John L. and Newgarden, Albert, eds. *Management Guides to Acquisitions.* New York: Wiley & Sons, 1969.

Hainer, Raymond M., Kingsbury, Sherman, and Gleicher, David B. *Uncertainty in Research, Management and New Product Development.* Cambridge, Mass.: Arthur D. Little.

Hainer, Raymond M., Kingsbury, Sherman, and Gleicher, David B., eds. *Uncertainty in Research, Management and New Product Development.* New York: Reinhold, 1967.

Hamberg, Daniel. *R&D/Essays on the Economics of Research & Development.* New York: Random House, 1963.

Hutchinson, G. Scott, ed. *The Business of Acquisitons and Mergers.* Montpelier, Vermont: Capital City Press, 1968.

Killeffer, David H. *How Did You Think of That?* Garden City, N.Y.: Doubleday, 1969.

Klaw, Spencer. *The New Brahmins.* New York: William Morrow, 1968.

Langhoff, Peter, et al. *Models, Measurement and Marketing.* Englewood Cliffs, N.J.: Prentice-Hall, 1965.

Laporte, Lowell. *R&D: Its Makeup, Management, and Measurement.* New York: The Conference Board, 1969.

Lazo, Hector and Corbin, Arnold. *Management in Marketing, Text and Cases.* New York: McGraw-Hill, 1961.

Lothrop, Warren C. *Management Uses of Research & Development.* New York: Harper & Row, 1964.

Mace, Myles L. *Management Problems of Corporate Acquisition.* Division of Research, Graduate School of Business Administration, Cambridge, Mass.: Harvard University, 1962.

Management of Research and Development. Paris: Organization for Economic Co-Operation and Development, 1971.

Managing Advancing Technology, Volume I. Strategies and Tactics of Product Innovation, Volume II. Creating an Action Team in R&D. New York: American Management Association, 1972.

Mansfield, Edwin. *The Economics of Technological Change.* New York: W. W. Norton, 1968.

——. *Industrial Research and Technological Innovation.* New York: W. W. Norton, 1968.

Mansfield, Edwin, Rapoport, John, Schnee, Jerome, Wagner, Samuel, and Hamburger, Michael. *Research and Innovation in the Modern Corporation.* New York: W. W. Norton, 1971.

McCarthy, George. *Acquisitions and Mergers.* New York: The Ronald Press, 1963.

McLoughlin, William G. *Fundamentals of Research Management.* New York: American Management Association, 1970.

Mergers and Markets. Studies in Business Economics No. 87. New York: National Industrial Conference Board, 1965.

Miller, Stanley S. *The Management Problems of Diversification.* New York: Wiley & Sons, 1963.

Morton, J. A. *Organizing for Innovation.* New York: McGraw-Hill, 1971.

Mueller, Robert Kirk. *The Innovation Ethic.* New York: American Management Association, 1971.

Myers, Sumner, and Marquis, Donald. *Successful Industrial Innovations.* National Science Foundations. Washington, D.C.: Government Printing Office, 1969.

Nemmers, Erwin E. *Managerial Economics.* New York, Wiley & Sons, 1962.

Perry, John H., ed. *Chemical Business Handbook.* New York: McGraw-Hill, 1954.

Phillips, C. F. and Duncan, O. J. *Marketing Principles and Methods.* Homewood, Ill.: Richard D. Irwin, 1959.

Risley, George. *Modern Industrial Marketing, A Decision-Making Approach.* New York: McGraw-Hill, 1972.

Roberts, Edward B. *The Dynamics of Research and Development.* New York: Harper & Row, 1964.

Saunders, J. H. *Careers in Industrial Research & Development.* New York: Marcel Dekker, 1974.

Scarf, Charles A. *Techniques for Buying, Selling, and Merging Businesses.* Englewood Cliffs, N.J.: Prentice-Hall, 1964.

Science, Technology and Innovation. Prepared for the National Science Foundation by Battelle-Columbus Laboratories. Washington, D.C.: National Science Foundation, 1973.

Shay, Philip W. *How to Get the Best Results from Management Consultants.*

Shennan, James G. *Case Studies in Diversification: Setting a Policy for Growth.* General Management Series No. 177: "Targets for Management," 1955.

Spitz, Eward A., ed. *Product Planning.* Auerbach, 1972.

Steger, Joseph A. *Factors in Innovation, Productivity, and Non-Productivity in Research: A Review and Pilot Study.* National Science Foundation, 1975. (Available from National Technical Information Service, U.S. Department of Commerce, Springfield, Virginia).

Steiner, George A. *Top Management Planning.* New York: Macmillan, 1969.

Success and Failure in Industrial Innovation. London: Science Policy Research Unit, Univ. of Sussex, Centre for the Study of Industrial Innovation, 1972.

Suits, C. Guy. *Suits: Speaking of Research.* New York: John Wiley, 1965.

Taton, R. *Reason & Chance in Scientific Discovery.* New York: Science Editions, 1962.

U.S. Department of Commerce. *Technological Innovation: Its Environment and Management.* Washington, D.C.: Government Printing Office, 1967.

Villers, Raymond. *Research and Development: Planning and Control.* New York: Financial Executives Research Foundation, 1964.

White, K. K. *Financing Company Expansion.* Research Study No. 64, 1964.

Wirt, John G., Lieberman, Arnold J., and Levien, Roger E. *R and D Management: Methods Used by Federal Agencies.* Prepared for Department of Health, Education, and Welfare by Rand Corporation. (Available from National Technical Information Service, U.S. Department of Commerce, Springfield, Virginia), 1974.

Articles

Ames, B. Charles. "Dilemma of Product/Market Management," *Harvard Business Review,* November-December 1971, p. 66.

Ames, B. Charles. "Marketing Planning for Industrial Products," *Harvard Business Review,* November-December 1968, p. 100.

Ansoff, H. Igor and Stewart, John M. "Strategies for a Technology-Based Business," *Harvard Business Review,* November-December 1967, p. 71.

Bjorksten, Johan. "Sure Guide to Unsuccessful Mergers." *Credit and Financial Management,* April 1965.

Brody, J. Kenneth. "Acquisitions and Mergers." *Trusts and Estates,* October 1966.

Cadbury, N. D. "When, Where and How to Test Market," *Harvard Business Review,* November-December 1975, p. 96.

"The Case Against Anti-Merger Policy." *Business Horizons,* Winter 1961.

Casenhiser, E. O. and Barkdull, C. W. "Planning for Acquisition." *Cost and Management,* March 1965.

Casler, L. A. "The Pitfalls of Acquisitions." *Dun's Review and Modern Industry,* June 1966.

"Changing Technology and the Future of Marketing." Conference Board Record, December 1974, pp. 22-26.

Christopher, William F. "Marketing Planning That Gets Things Done." *Harvard Business Review,* November-December 1970, p. 56.

Clewett, Richard M. and Stasch, Stanley F. "Shifting Role for the Product Manager." *Harvard Business Review,* November-December 1975, p. 65.

Cochran, Stuart W. "Mergers–The Easy Road to Growth." *Commerce,* March 1962.

"The Commercialization of Internally Supported Resources." Conference Board Report 695, 1976.

"Consolidation vs. Combination." *The Accounting Review,* January 1962.

"Corporate Acquisition: A Case Study." *New York Certified Public Accountant,* July 1960.

"Corporate Merger Counsel." *Management Controls,* April 1963.

Cowperthwaite, Gordon H. "Administering a Reorganization." *Management Controls,* July 1962.

"Distinguishing Between Purchase and Pooling." *The Journal of Accountancy,* June 1961.

Davis, Richard E. "Compatibility in Corporate Marriages." *Harvard Business Review,* November-December 1968, p. 86.

Eastlack, Joseph O., Jr. and McDonald, Philip R. "CEO's Role in Corporate Growth, (Problems in Review)," *Harvard Business Review,* November-December 1970, p. 150.

"Elementary Tax Concepts of Combining and Reorganizing Corporations." *New York Certified Public Accountant,* December 1961.

Flanagan, Terence N. "Hidden Dangers in Mergers." *Credit and Financial Management,* May 1962.

Folz, David F., and Weston J. Fred. "Looking Ahead in Evaluating Proposed Merger." *N.A.A. Bulletin,* April 1962.

Gladstone, William L. "Tax Aspects of the Allocation of Purchase Price of a Business." *The Journal of Accountancy,* October 1966.

"A Good Look at Goodwill in Corporate Acquisitions." *The Banking Law Journal,* December 1961.

Greene, Mark R. "How to Rationalize Your Marketing Risks." *Harvard Business Review,* November-December 1969, p. 114.

"Guideposts for Evaluating a Prospective Acquisition." *Budgeting,* January 1964.

Hanan, Mack. "Corporate Growth Through Internal Spin-Outs." *Harvard Business Review,* November-December 1969, p. 55.

——. "Corporate Growth Through Venture Management." *Harvard Business Review,* November-December 1969, p. 43.

Handler, Milton, and Robinson, Stanley D. "The Supreme Court vs Corporate Mergers." *Fortune,* January 1965.

Handy, John L. "How to Face Being Taken Over." *Harvard Business Review,* November-December 1969, p. 109.

Holmes, Melvin C. "Buyer's Counsel Looks at Acquisitions." *The Business Lawyer,* April 1963.

Hopper, W. J. "Take-Overs and Mergers." *Management,* May 1962.

"How Not to Sell Your Company." *Harvard Business Review,* May-June 1961.

"How to Evaluate a Potential Acquisition." *Business Management,* August 1966.

"How to Grow Fast by Merger." *Business Management,* February 1962.

"Impact of Merger Accounting on Post-merger Financial Reports." *Management Accounting,* December 1965.

"Industry Turns to Convertible Preferred Stock Issues to Finance Acquisitions." *Chem. Eng. News,* August 22, 1966.

Kelsey, John P. "Carry-overs in Corporate Acquisitions." *Taxes,* March 1962.

Kinard, Hargett Y. "Financing Mergers and Acquisitions," *Financial Executive,* August 1963.

Lang, H. J. "The Human Factor in Mergers and Acquisitions." *Management Record (NICB),* January 1963.

Leavitt, Theodore. "New Markets, The-Think Before You Leap." *Harvard Business Review,* November-December 1969, p. 53.

Lee, Ralph E., Jr. "Acquisition of a Business–Accounting and Financial Aspects." *Taxes,* March 1962.

Leeper, D. S. "Acquisition–Know What You're Getting." *The Controller,* August 1962.

Leighton, Charles M. and Tod, Robert. "After the Acquisition: Continuing Challenge." *Harvard Business Review,* November-December 1969, p. 90.

Link, Richard M. and Wheat, Francis M. "Financing and Legal Aspects of Acquisitions," in "Functions of Corporate Legal Departments." Berkeley, California: Continuing Education of the Bar of the State of California, University of California, 1961.

Linowes, David F. "Is Merger Necessary?" *Advanced Management Journal,* April 1965.

Love, George H. "Some Questions About Mergers." *Vital Speeches of the Day,* October 15, 1965.

Martin, David Dale. "The Brown Shoe Case and the New Antimerger Policy." *The American Economic Association,* June 1963.

Mather, C. L. "Corporate Acquisitions–Tax Accounting Consequences." *N.A.A. Bulletin,* March 1965.

McLaren, Robert. "Mergers and Acquisitions: The New Direction." *Public Relations Journal,* September 1965.

"Merger an Art and a Science." *The Controller,* July 1961.

"Mergers and Market Size." *Business Record,* New York: National Industrial Conference Board, April and July 1959.

Messing, Richard F. "R & D Evaluation Important in Acquisitions." *Chem. Eng. News,* April 20, 1964.

"New Emphases in Marketing Strategies." Conference Board Record, August 1976, p. 35-39.

"New-Product Development." Conference Board Record, September 1976, pp. 25-28.

"New-Product Pressures." Conference Board Record, June 1971, pp. 16-24.

"New-Product Risk in Industrial Markets." Conference Board Record, November 1967, pp. 26-31.

Patton, E. J. "Tax Implications of Mergers and Acquisitions." *Financial Executive,* January 1963.

"The Perilous Quest for Acquisitions." *Dun's Review and Modern Industry,* July 1965.

Peterson, Russell W. "New Venture Management in a Large Company." *Harvard Business Review,* November-December 1967, p. 68.

"Picking a Merger Partner: Some Guidelines." *Management Controls,* April 1963.

"The Place of Scientific Techniques in Mergers and Acquisitions." *The Controller,* July 1961.

"Pre-Decision Reporting on Acquisition of a Plant or Company." *N.A.A. Bulletin,* January 1961.

"Premeditated Merger." *Harvard Business Review,* January-February 1961.

Prince, Robert M. "Overseas Operations: How to Make an Acquisiton You Can Live With." *The Management Review,* November 1965.

"Problems in Combining Reorganizations." *Taxes,* November 1961.

"The Product Manager System." *Experiences in Marketing Management 8,* 1965.

Pryor, Millard H., Jr. "Anatomy of a Merger." *Michigan Business Review,* July 1964.

Rockwell, Willard F., Jr. "How to Acquire a Company." *Harvard Business Review,* November-December 1968, p. 121.

Rowe, Frederick M. "Trends in Mergers and the Law," in "Public Policy toward Competition: New Antitrust Directions." *The Conference Recorder* (An Occasional Series), New York: National Industrial Conference Board, 1962.

Sapienza, Samuel R. "Pooling Theory and Practice in Business Combinations." *The Accounting Review,* April 1962.

Schaffer, Robert H. "Putting Action Into Planning." *Harvard Business Review,* November-December 1967, p. 156.

Shad, John S. R. "The Financial Realities of Mergers." *Harvard Business Review,* November-December 1969, p. 133.

Smalter, Donald J. and Lancey, Roderic C. "P/E Analysis in Acquisition Strategy." *Harvard Business Review,* November-December 1964.

Smith, A. W. "Petroleum Mergers Result in Organizational Change." *Industrial Management,* January 1965.

Stacy, Nicholas H. "Are Mergers Misunderstood?" *The Manager,* June 1963.

Stacy, Nicholas H. "Mergers: Facing the Facts." *The Manager,* June 1965.

Stephan, Edmund A. "Acquisition Trouble Stops." *The Business Lawyer,* January 1966.

Stern, Louis W. "Mergers Under Scrutiny (Keeping Informed." *Harvard Business Review,* November-December 1969, p. 18.

"Thinking of Merging? Here Are Some Tips." *Steel,* July 19, 1965.

Tincher, William R. "Yardsticks for Evaluating Company Acquisitions." *The Management Review,* October 1964.

"Transfer of Business through Tax-free Reorganization." *The Monthly Digest of Tax Articles,* July 1961.

Vance, Jack O. "Is Your Company a Take-Over Target?" *Harvard Business Review,* November-December 1970, p. 98.

Vancil, Richard F. "So You're Going to Have a Planning Department." *Harvard Business Review,* November-December 1967, p. 88.

Wakefield, B. R. "Mergers and Acquisitions." *Harvard Business Review,* September-October 1965.

Weiner, Jack B. "The Big Management Shake-up." *Dun's Review and Modern Industry,* June 1964.

Weston, J. Fred. "Planning for Corporate Merger." *California Management Review,* Spring 1963.

"When a Company Shops for a Company." *Research for Industry,* March-April 1964.

"What's An Acquisition Worth?" *Dun's Review and Modern Industry,* March 1964.

"What to Look for When You Buy a Company." *Industrial Marketing,* May 1962.

White, George R. and Graham, Margaret B. W. "How to Spot a Technological Winner." *Harvard Business Review, March-April 1978, pp. 146-152.*

Symposia

Alstadt, Donald M., President, Lord Corporation. "Management of Change." Change Unlimited–How to Cope, March 1975.

Avery, Henry J. William McNeil. "Strategy Planning for Growth." Planning for Profitable Growth, October 1970.

Black, Paul B. "How to be an Effective Planner." Planning for Profitable Growth, October 1970.

Blake, Edmund J., Vice President, Pioneer Lands Corporation. "Venture Capital and Corporate Development." Changing Times in Commercial Development, March 1971.

Coffin, David L., The Dexter Corporation. "Opportunity and Risk in Commercial Development." Opportunity and Risk in Commercial Development, 33d annual meeting, March 1977.

Collier, James R. "Effective Long-Range Business Planning." Planning for Profitable Growth, October 1970.

Collins, David E., General Counsel, Johnson & Johnson. "Industry/Government Interface–Causes and Consequences." The Federal Government/Industry Interface, September 1975.

Conn, Doak O., III, E. I. du Pont. "Commercialization of Technology Developed to Meet Internal Needs." Short Term Profits or Long Term Growth, fall meeting, September 1976.

Dempsey, James F., Air Products. "Strategic Planning at Air Products." Chemical Marketing, and Economics, American Chemical Society, August 1977.

Graziano, Frank J., President, Crompton & Knowles Corporation. "Commercial Development Abroad to Maximize Profits at Home." Changing Times in Commercial Development, March 1971.

Harbison, Earle H., Jr. Monsanto Company. "How Do You Get Commitment to Growth?" Short Term Profits or Long Term Growth, fall meeting, September 1976.

Hout, Thomas M., Boston Consulting Group. "The Economics of International Competition." Short Term Profits or Long-Term Growth, fall meeting, September 1976.

Joy, Paul W., Vice President, The Carborundum Company. "New Business Development." Evaluation of New Projects and Successful Techniques to Implement Them, November 1972.

Nobis, John F., American Cynamid Corporation. "Acquisition Vs. R&D Vs. Technology Purchase." Short Term Profits or Long Term Growth, fall meeting, September 1976.

Poper, Frank J., Director of Economic Analysis, General Motors Corporation. "Responding to External Change in the Automotive Market." Change Unlimited–How to Cope, March 1975.

Reiser, Robert F., Manager, Xerox Corporation. "Commercial Development Under Fire." Changing Times in Commercial Development, March 1971.

Roher, William C., Gulf Oil Chemicals Company. "The Use of Decision Analysis in Commercial Development." Opportunity and Risk in Commercial Development, 33d annual meeting, March 1977.

Throdahl, Monte C. "Corporate Objectives and Growth." Planning for Profitable Growth, October 1970.

Wilson, Ian H. "Corporate Development in a Changing Society." Planning for Profitable Growth, October 1970.

INDEX

Acquisition, growth, by, 249
Action objectives in the communications program, 169
Advertising (*see* Communications program)
Analytical skills, 26
Annual reports, 199-200
Annual Survey of Manufactures, 194
Antitrust legislation, in environment of new business ventures, 106
Applications research, 5-6
Applied Science and Technology Index, 201
Attitudinal objectives in the communications program, 170-172

Bibliographies, 201
Business Conditions, 200
Business cycles, in environment of new business ventures, 103
Business enterprise, 112
Business in Brief, 200
Business Periodical Index, 201
Business planning, 43-60
    alternatives, 52-54
        defining, 53
        evaluating, 54
        identifying, 52-53
        selecting, 54
        structuring, 53
[Business planning]
    the business planning group, 51
    commercial development and, 55-57
    the computer in, 50
    defined, 44
    the environment and, 49
    formal, 47
    the future and, 44
    history of, 44-46
    informal, 47
    master strategy, 52
    need for, 46
    origins of, 44
    pacing, 56-57
    performance areas and, 47-48
    as a process, 47
    profits and, 50
    purpose of, 47
    scope of, 48-49
    sequencing, 56-57
    timing, 56-57
Business Publications Edition of Standard Rate and Data Service, 160
Business Review, 200
Business Service Checklist, 194
Business Statistics, 195
Buyer's costs, 152

Carbide and Carbon Chemical Company, 23
Case histories, 253-291

[Case histories]
E. I. Du Pont de Nemours & Co., 264-273
General Electric Company, 281-288
Loctite Corporation, 253-264
PPG Industries, 273-281
CCDA Honor Award, 23
Census of Manufactures, 194
Chemical and Rubber Industry Report, 195
Chemical industry, 7-8
Chemical Marketing Research Association, 9
Chemical Market Research in Practice, 76
Commercial development, 3-27
business planning and, 55-57
departments, 8-9
environmental pressures and, 210
external sources of assistance, 181-203
consultants, 181-188
government sources, 192-198
industrial research institutes, 188-189
literature sources, 198-202
philanthropic organizations, 190
professional associations, 190-192
trade associations, 190-192
universities, 189
finance and, 223-227
the future, 12-13, 213-215, 245-250
acquisitions, 249
changing nature of commercial development, 246-248
increasing cost of research and development, 248-249
increasing economies of scale, 248
lower profit margins, 249-250
mergers, 249
technological change, 248
history of, 3-7
inputs to, 75-95
[Commercial development]
classification of major research areas, 76-79
market development, 82-87
marketing research, nature and scope of, 75-76
product research, 80-82
scope of technical service activities, 87-92
service research, 80-82
international, 10-11, 104, 231-243
export sales, 233-234
joint ventures, 238-240
licensing, 236-238
marketing channels, 234-236
new product testing, 241-242
trade negotiations, 104
wholly owned subsidiary, 240-241
interrelationships, 217-227
management and, 11, 207-216
interrelationships, 218-220
marketing and interrelationships, 222-223
organizing and managing, 15-32
enabling skills, changes in, 26
ideal, 16
locations by levels, 17-21
new product committees, 21-22, 31-32
types of effort, changes in, 24-25
present situation, 7-12
production and interrelationships, 222
purchasing and, 223-227
research and development and, interrelationships, 220-221
responsibility for, 211-213
role of, 207
study of attitudes toward, 207-210
Commercial Development Association (CDA), 4, 6-9, 10, 12, 15
Conference Board, 17-22, 35-36
on consultant selection, 184-185
tables, 28-32

Communications program, 157-180
  budget for, 174-175
  demonstrations, 166-167
  direct mail, 163, 164
  editor contact, 163
  electronic advertising, 159-161
  exhibits, 165-166
  inquiry handling, 167
  list development, 164
  literature, 164-165
  media mix, 175
  movies, 167
  news release, 162
  objectives, 168-172
    action, 169
    attitudinal, 170-172
    educational, 169-170
  package preparation, 164
  parameters for, 172-173
  positioning, 174
  press conference, 162-163
  proper sequence for, 176-180
  publicity, 161-162
  responsibility for, 168-175
  seminars, 166-167
  significance of, 157-158
  space advertising, 159-161
  techniques, 158-167
  timing, 173-174
Competition:
  ideas from, 39
  new product and, 39-40, 223
  pricing and, 144, 150-152
Computers, in business planning, 50
Concept testing, new projects, 122-125
Congressional reports, 200-201
Consultants, 181-188
  advantages of using, 182-183
    anonymity, 182
    cost savings, 182
    imaginative ideas, 183
    improved contacts, 183
    objectivity, 182-183
  liaison between marketing research and, 90
  for new products, 37
[Consultants]
  selecting, 183-188
    Conference Board on, 184-185
  work appraisal, 188
Consultants and Consulting Organizations Directory, 184
Consumer Product Safety Commission, 193
Controls of legislation, in environment of new business ventures, 105
Corporate level organization, 16-18
Cost:
  buyer's costs, 152
  consultants and, 182
  new product, 145-146, 151-152
  of research and development, future increase of, 248-249
Court records, 200-201
Critical path (PERT) technique, 176-180
Current Industrial Reports, 194-195
Customers:
  suggestions from, 36
  technical needs of, 88

Decision point evaluation procedure, 111
Defendability factor, 139-140
Demographic forces, in environment of new business ventures, 104
Demonstrations, 166-167
Departments, new venture management, 67, 71
Direct mail, 163, 164
Direct Mail List Edition of Standard Rate and Data Service, 164
Directories, 201
Directory, Association of Management Consultants, 184
Directory of Membership and Services, 184
Distribution, 77
Distributors, suggestions from, 36-37
Divisional level organization, 16-18

Domestic International Sales Company (DISC), 236
Dow Chemical Company, 24-26

E. I. Du Pont de Nemours & Co., 61-62
  case history, 264-273
Economic factors, in environment of new business ventures, 103-104
Economies of scale, increasing, 248
Editor contact, 163
Educational objectives in the communications program, 169-170
Electronic advertising, 159-161
Encyclopedia of Associations, 190
Energy Standing Order Service, 196
Engelhard Minerals and Chemicals Corporation, 35
Engineering Index, The, 201
Entrepreneurial management, 64-66
Entreprenuership and the Corporation, 61
Environmental Protection Agency (EPA), 193
European Guide to Industrial Marketing Consultancy, 184
Exclusion charts, 154
Exhibits, 165-166
Export sales, 233-234

Federal Energy Administration, 193
Federal Statistical Directory, The, 192, 201
Finance:
  commercial development and, 223-227
  new projects and, 118-119, 224
Financial institutions, as literature sources, 200
Fixed bench marks, 107-109
Floor price, 152
Freedom of Information Act, 193
Future aspects:
  business planning and, 44
  commercial development, 12-13, 213-215, 245-250
    acquisitions, 249
    changing nature of commercial development, 246-248
    increasing cost of research and development, 248-249
    increasing economies of scale, 248
    lower profit margins, 249-250
    mergers, 249
    technological change, 248
  of management, 213-215

General Electric Company, 68-70
  case history, 281-288
Government Inventions for Licensing, 197
Government regulations, as a basis for growth, 26
Government Reports Abstracts, 195
Government sources for commercial development, 192-198
Guide to American Directories, 201
Guide to U.S. Government Serials and Periodicals–Documents Index, 201

Harvard Business Review, 226
House organs, 199
How to Get the Best Results from Management Consultants, 186

Indexes, 201
Industrial research institutes, 188-189
Inquiry handling, 167
International commercial development (*see* Commercial development, international)

International Convention for the Protection of Industrial Property, 237
International Directory of Marketing Research Houses and Services, 184
Inventors, 39
Investment:
- new projects, 136
- return on, 128

Joint ventures, international, 238-240

Labor, in environment of new business ventures, 105-106
Licensing, 40
- international, 236-238

List development, 164
List price, 77-78
Literature:
- for communications program, 164-165
- for new products, 164-165
- as sources for commercial development, 198-202

Loctite Corporation (case history), 253-264

Management (*see also* New-venture management):
- commercial development and, 11, 207-216
  - interrelationships, 218-220
  - study of attitudes toward, 207-210
- communications with, 219-220
- defined, 62
- entrepreneurial, 64-66
- future of, 213-215
- responsibility of, 211
  - new products, 30
- scientific, 64-65
- use of, for existing products, 29

Market development, 4-5, 10
- criteria for, 116-118
- function of, 82-87

Marketing:
- characteristics of, 77
- commercial development and interrelationships, 222-223
- international, 234-236
- organization, 18
- test, 81-82

Marketing communications (*see* Communications program)
Marketing Information Guide, 194
Marketing of Chemical Products, The, 61
Marketing research, 35
- agencies, 37
- liaison between technical consultants, 90
- nature and scope of, 75-76

Master strategy plan, 52
Materials suppliers, suggestions from, 37
Mathematical models, in new project screening, 125-129
Media mix, 175
Mergers, future growth by, 249
Monetary policy, in environment of new business ventures, 104
Monthly Catalog of Government Publications, 193
Monthly Catalog of United States Government Publications, 201
Monthly Energy Review, 196
Monthly Letter, 200
Monthly Review, 200
Movies, 167
Moving bench marks, 109-112
Multinational company, defined, 232

Nationalization of industries, in environment of new business ventures, 105
National Science Foundation, 39

National Technical Information Service (NTIS), 193, 195-197
National Trade and Professional Associations of the United States and Labor Unions, 190
New activities objectives (NAO), 129-130
New business ventures, 20, 30-31
  defined, 62
  environment of, 99-114
    antitrust legislation, 106
    business cycles, 103
    business enterprise, 112
    controls legislation, 105
    demographic forces, 104
    economic factors, 103-104
    evaluation procedure, 101
    external factors, 103-106
    financial results of business, 108-109
    fixed bench marks, 107-109
    growth of business, 108
    internal factors, 107-112
    international trade negotiations, 104
    kind of business, 107
    labor, 105-106
    long-term trends, 101-103
    monetary policy, 104
    moving bench marks, 109-112
    nationalization of industries, 105
    opportunity categories, 100
    orientation of business, 107-108
    political factors, 105-106
    pressure groups, 104-105
    selection procedure, 101
    size of business, 107
    social factors, 104-105
    tariffs, 104
    taxes, 103-104
    technical factors, 106
New products:
  committees, 21-22
    use of, 31-32
  competition and, 39-40, 223
[New products]
  consultants for, 37
  criteria for, 115
  development, 89
  for established markets, 78-79
  evaluation of, 80-81
  format for report on, 219
  innovation and creation of ideas for, 33-41
    external sources for, 36-39
    monitoring competitive activities, 39-40
    origins of, 33-36
  literature for, 164-165
  location of activities, 28-29
  organization, 19-21
  pricing, 143-156
    competitive considerations, 144, 150-152
    cost considerations, 145-146, 151-152
    defining objectives, 147-148
    exclusion charts, 154
    factors affecting, 143-147
    market expectations in, 144
    marketing factors, 145
    product characteristics in, 145
    profit considerations, 145-146
    sales expectations in, 144
    strategies for, 148-151
  responsibility of management, 30
  testing, international, 241-242
New projects, selection and evaluation of, 115-142
  basic relationship, 132-134
  concept testing, 122-125
  defendability factor, 139-140
  evaluation form, 120-121
  financial criteria, 118-119, 224
  investment factor, 136
  market criteria, 116-118
  mathematical models, 125-129
  new activity objectives, 129-130
  number of projects, 131-132
  product criteria, 115-116
  project champion, 140
  project life factor, 135-136

[New projects, selection and evaluation of]
proposal checklists, 122
prototype evaluation, 122-125
risk factor, 136-139
slavage value factor, 136
selling price factor, 135
size of project, 130-131
statistical screening, 125-129
unit cost factor, 135
volume factor, 134-135
NEWS (New Product Early Warning System), 128
News releases, 162
New venture management, 61-74
defining terms, 62
entrepreneurial, 64-66
management as a factor in success, 62-64
routes to, 66-70
department for, 67, 71
spin-offs, 68-71
task forces, 68, 71
venture capital organizations, 68
scientific, 64-65
New York Times Index, The, 201

Occupational Safety Hazard Administration (OSHA), 193
Options in New-Product Organization, 17

Package preparation (direct mail), 164
Packaging, 77
Patent attorneys, 38
Patent brokers, 39-40
Patents, 39-40
foreign, 237-238
surplus, 40
Penetration pricing, 149
Performance areas, business planning and, 47-48
PERT (critical path) technique, 176-180
Petroleum Situation, The, 200
Petroleum Statistics Report, 196
Philanthropic organizations, 190
Planning (*see* Business planning)
Political factors, in environment of new business ventures, 105-106
Positioning in the communications program, 174
PPG Industries (case history), 273-281
Press conferences, 162-163
Press kits, 163
Pressure groups, in environment of new business ventures, 104-105
Pricing, 77-78
floor price, 152
list price, 77-78
new products, 143-156
competitive considerations, 144, 150-152
cost considerations, 145-146, 151-152
defining objectives, 147-148
exclusion charts, 154
factors affecting, 143-147
market expectations in, 144
marketing factors, 145
product characteristics in, 145
profit considerations, 145-146
sales expectations in, 144
strategies for, 148-151
penetration, 149
skimming, 149
Product bulletins, 199
Product designers, 38-39
Products, new (*see* New products)
Product specialists, retired, 38
Product testing, 26
international, 241-242
Professional associations, 190-192
Profits:
business planning and, 50
lower future margins of, 249-250
new products, 145-146
Project life factor, 135-136

Project prognosis guide, 227
Projects, new (*see* New projects, selection and evaluation of)
Project teams, 20, 30
Proposal checklists, new projects, 122
Prospectuses, 200
Prototype evaluation, new projects, 122-125
Proxy statements, 200
Public Affairs Information Service Bulletin, 201
Publicity, 161-162
Published Services, 196
Purchasing, commercial development and, 223-227

Quality level, 172-173

Regional reports, 202
Report on Western Europe, 200
Research:
  applications, 5-6
  classification of major areas of, 76-79
  marketing, 35
    agencies, 37
    liaison between technical consultants and, 90
    nature and scope of, 75-76
  on services, 80-82
Research and development (R&D), 35
  breakdown of expenditures, 39
  commercial development and interrelationships, 220-221
  increasing cost of, 248-249
  organization, 18-19
Resellers, suggestions from, 36-37
Responsibility:
  for commercial development, 211-213
  communications program, 168-175
  of management, 211
    new products, 30
Return on investment (ROI), 118-119
Risk:
  analysis chart, 126
  new projects, 136-139

Sales:
  export, 233-234
  new product expectations, 144
Sales department, services to the, 89
Salvage value factor, 136
Scientific managmenet, 64-65
Selected Water Resources Abstracts Journal and Index, 196
Selling price factor, new projects, 135
Seminars, 166-167
Skimming pricing, 149
Social factors, in environment of new business ventures, 104-105
Space advertising, 159-161
Special Foreign Currency Science Information Program (SFCSIP), 197
Specifications, 26, 77
Specifications of Profits with Interaction under Trial and Error Response (SPRINTER), 128
Spin-offs, new-venture management, 68-71
Standard Industrial Classification (SIC), 160
Standardized tests, 26
Standard Rate and Data Service, Inc., 201
Statistical Abstract of the United States, 195
Statistical models, new project screening, 125-129
Subsidiary, wholly owned, 240-241
Successful Commercial Chemical Development (Corley), 23
Survey of Current Business, 195

Tariffs, 247
  in environment of new business ventures, 104

Taxes, in environment of new business ventures, 103-104
Technical factors, in environment of new business ventures, 106
Technical Help to Exporters (THE), 197
Technical service activities:
 current situation, 91-92
 scope of, 87-92
Technical Service Association, 4, 6
Techniques in Financial Analysis, 108
Technology:
 forecasting, 49
 increasing rate of change in, 248
 keeping abreast of, 89
Tech Progress Reports, 197
10K reports, 199-200
Test marketing, 81-82
Think tanks, 210-211
Timing, in the communications program, 173-174
Toxicological skills, 26
Trade associations, 190-192
Trade journals, 199

Union Carbide Corporation, 23-24
Unit cost factor, new projects, 135
United States Bureau of the Census, 194-195
United States Department of Commerce, 39, 194-198
United States Government Manual, 192
Universities as external sources of assistance, 189

Venture capital organizations, 38, 68, 71
Videotape, 167
Volume factor, new projects, 134-135

Wall Street Journal Index, The, 201
Wars, in environment of new business ventures, 104
Waste circulation, 159
Weekly Government Abstracts, 195
Weekly Petroluem Statistics Report, 196
Wholesalers, suggestions from, 36